ALGAE TO THE RESCUE!

Aphanizomenon flos-aquae (AFA)
is the wild blue-green algae that is the most
nutrient dense food available on planet Earth.

In this lively book, veteran educator, chemist,
and author Karl J. Abrams carefully and clearly
explains just how this natural super-food from
majestic Upper Klamath Lake, Oregon,
can work for you.

Here is a book that galvanizes
the information about the super-food
you have been seeking. It's time to restore
the bounce in your step and achieve the health
and vitality you deserve!

ALSO BY KARL J. ABRAMS

A Course in Experimental Chemistry, I & II

FREEMAN COOPER & CO., PUBLISHERS

All "A"s with Algae

Energize Your Mind with Nutritional Blue-Green Algae

COMING SOON FROM LOGAN HOUSE PUBLICATIONS

SPRING 1997 • ISBN 1-889152-01-3

Logan House PUBLICATIONS

1996

Algae to the Rescue!

Everything you need to know about nutritional blue-green algae

KARL J. ABRAMS

FOREWORD BY

Paul Swanson, M.D.

INTRODUCTION BY

Kathleen De Remer, M.D.

ISBN 1-889152-00-5
Library of Congress Catalogue Card Number:
96-076282

Manufactured in Canada.

PRINTING HISTORY
First edition: AUGUST, 1996
Second edition: OCTOBER, 1996
2 3 4 5 6 7 8 • 00 99 98 97

EDITOR
Sandra Rush

BOOK DESIGN
Burch Typografica

ILLUSTRATIONS
Michael Sumner

COVER DESIGN
Andrews/Keys Associates

PHOTO CREDITS
Front cover: David Collins
Back cover: T. W. Brown

Logan House PUBLICATIONS

4307 BABCOCK AVENUE
STUDIO CITY · CALIFORNIA 91604
818.763.0405 · 763.0402 *fax*

Leonard Buschel, PUBLISHER

DEDICATION

Dedicated to Earl Ettienne, Ph.D.
Friend, microbiologist, and nutritionist,
who lived his life in hope of a wider celebration

"Feeling is the language of the soul
and
heart connections
are the nutrients of the soul."

– DARYL J. KOLLMAN
San Francisco,
February 23, 1996

ACKNOWLEDGMENTS

Many thanks to my teachers, students, peers, and friends

for helping to make this book a reality.

Philip Andrews • Maria Curtis • Jerry Milgram
David Fishman • Dave Rathbun • Nancy Fern • Jeffrey Bruno
Jeffrey Anson • Dallas Saunders • Steven Dubin • Kathleen De Remer
Matt Damsker • Sandra Rush • Michael Whyte • Noel Fox
Eric Rappaport • Bill Webb • Melissa & Lala
Sally Abrams • William Abrams
Benjamin Buschel • Alex Kevorkian • Eric Sherman
Syris Falken • Linus Pauling • Tony Kent • Mark Segars
Bruce Buschel • Scott Springer • Daryl Kollman • Jacyra Rawlins
John Robbins • Herman Aihara • Michio Kushi • Gerry Thompson
Stewart Kenter • Kathleen Burch • Robert Downey, Sr.
Tim Brown • David Miller • David Collins
Dr. Paul Swanson • Torkom Saraydarian

I would especially like to acknowledge

the indefatigable inspiration of my publisher Leonard L. Buschel,

his boundless optimism, faith, and absolute commitment to this project.

Overview

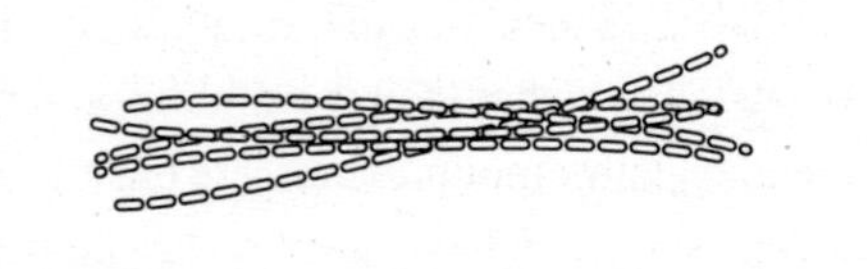

DR. KATHLEEN DE REMER, M.D.

Introduction

WHEN I WENT TO MEDICAL SCHOOL, WE WERE TAUGHT THAT if you ate well – and it was tacitly assumed that most people did – then you didn't need any vitamins or other form of supplementation. We now know that less than 10 percent of the population of the United States even meets the RDA standards, which were established to prevent certain deficiency diseases. Our soil was noted by the late 1940s to be nearly devoid of nutrients; fruits and vegetables have been hybridized to extend shelf-life at the sacrifice of natural enzymes and vitamins; pesticides and herbicides are a way of agricultural production; hormones and antibiotics fed to the animals find their way to our tables. Dietary consumption of fat is close to 40 percent, much higher than is healthful, and fiber intake is far lower than the recommended daily amount. These are but a few of the sobering details of modernization that suggest that the nutrition teachings of my professors are obsolete.

In addition, science has come to appreciate in the past decade or so that oxygen which is essential for life, is also extremely toxic because when we metabolize oxygen, free radicals are generated. Free radicals are unstable molecules that are missing an electron. In a predictable fashion they steal electrons from cells in the body. This process, repeated over and over again destroys the normal cells, interferes with cell function and memory, and creates more free radicals. Thousands of these reactions are created by just one free radical. The sun, chemical additives, pesticides, smoking, a high-fat diet, and overexercise are just some of the contributing factors leading to the generation of free radicals. Why is this important? At least 30 malignancies, many degenerative diseases, much of the cardiovascular disease, immune dysfunction, and the debility and deterioration we have come to associate with aging are due to free radical damage. Conservatively, it has been suggested that 50 to 60 percent of the heart disease, cancer, and declines due to "aging" can be prevented with lifestyle change, improved nutrition, and the addition of antioxidants.

We can't avoid free radicals, but we can minimize their production and we can strive to inactivate them. Antioxidants are stable molecules that act as electron donors, thereby stabilizing the free radicals. They, in turn, do not become free radicals. We have been provided with our own antioxidant defense system, superoxide dismutase, but this is simply not enough. Recent scientific literature has identified nutrients, vitamins, and phytochemicals that have the ability to interfere with this cascade of molecular chaos created by free radicals.

Wise men and women have known all along that nutrition is an essential component to our health and well being. The medical community has been woefully behind in acknowledging this. With the soaring costs of health care and the growing fund of nutritional research, there is now a credible movement toward prevention.

The scientific literature that I have reviewed and the knowledgeable people with whom I have communicated provide compelling data on the role of antioxidants, phytochemicals, and proper nutrition in health. Maybe all the evidence is not in yet, but I would rather not wait to begin a program I feel will contribute to wholeness and health. I have modified my diet, for example, by decreasing my fat intake and increasing fiber, and I also take antioxidants.

What is interesting is that my search for antioxidant supplementation for horses led to blue-green algae for myself. I had come close to losing a prize quarter horse mare. Hospitalization at an excellent equine facility provided no tangible cause or treatment to her growing debility. If human nutrition is so poor, what has domestication, antibiotics, worming, confining, agricultural practice, and the processing of food done to our animals? I reasoned that if antioxidants and other supplements could make a difference in humans, certainly other animals could benefit as well. But finding a product for animals was no easy task. Serendipitously, a friend suggested that I speak with one Professor Karl J. Abrams about blue-green algae. With a certain amount of skepticism I listened and learned. With great fascination I read the first draft of this manuscript and came to truly appreciate the natural biological wonders of this simple algae.

Because of my educational background, I need to see the data, the scientific development and reasoning. Stories and anecdotal reports aren't enough – I want substance. The blue-green algae is a simple organism that contains an elaborate warehouse of amino acids, enzymes, vitamins, minerals, antioxidants, and other micronutrients with incredibly complex functions. With the scrutiny of a scientist and the skill of a teacher accustomed to communicating detail and transforming what is for some students a tedious dry experience in science into an exciting adventure, Karl Abrams shares with us his research into the biochemical and biological properties of this simple one-celled algae. I trust that you as a reader will enjoy this magical molecular journey and too will understand that nutrition – rather, adequate nourishment – is essential to the process of healthful living.

Dr. Kathleen De Remer, M.D.
KAYSVILLE, UTAH
JUNE 1996

KATHLEEN DE REMER, M.D., *was born in Princeton, New Jersey, in 1947. She graduated with Honors from Jackson College of Tufts University where she had designed an independent study progam on human genetics and aging. Kathleen received her Medical Doctorate in 1975 from the Robert Woods Johnson Medical School and practices internal medicine in Utah. Kathleen has also actively participated in the Multi-Disciplinary Oncology Program at Huntington Memorial Hospital in Pasadena, California. She has been awarded, in 1984, the "AMA Physician's Recognition Award" and "Foremost Women of the Twentieth Century" award.*

DR. PAUL SWANSON, M.D.

Foreword

TODAY THE WORLD IS AT AN EVOLUTIONARY CROSSROADS. To help proceed in the best direction, let us look into our origin. Seven million years ago a primate species began to evolve into upright-walking early humans. Living in tribes and eating wild foods, they continued to evolve toward our present form. Those humans possessed a vigor of body and keenness of sense matching animals in the wild, along with precisely tuned and evolving discernment, understanding, communication, adaptability, creativity, hope, reverence, and courage. Humans in that long ago era were noble in body, heart, and mind, and were profoundly healthy.

Health is a highly tuned resonance of the myriad processes of a living organism with a specific environment. We've lost sight of the awe-inspiring nature of our existence partly because we've lost the edge on the nearly perfect resonance of health that we had during our evolution. We've lost this edge because our environment and condition have changed. Our genetics have been set in place over four billion years, and as humans over seven million years; our biological nature has not changed much in the last few thousand.

One of the greatest changes in the conditions in which humans live has been in what we eat. Humans no longer eat much raw wild food. The development of agriculture, and now industrial farming, have seriously worsened the quality of the food we eat.

Humans don't recognize that the edge of resonant health has been lost because there is no example of resounding health, no point of reference, in daily life. Everyone has been equally dulled and damaged. In the past fifty years this dulling of the resonant edge of health has progressed to almost universal degenerative disease and a general weakness, listlessness, and apathy.

We are at a turning point in history, a crisis that is both a danger and an opportunity. Now, at this time in history, comes Algae to the Rescue!

The native blue-green algae growing wild and abundantly in Oregon's Upper Klamath Lake, one of the world's most volcanic mineral-rich and unpolluted lakes, provides an opportunity for humans for something closer to an ancient health.

Those who eat this algae and gain understanding of a higher human health – the sharp edge of human health that has been lost – have a new responsibility. That responsibility is first to learn, then to teach.

Much thanks to Karl Abrams for this excellent book. It is a significant step ahead toward the great task of learning and teaching.

Dr. Paul Swanson, M.D.
GRASS VALLEY, CALIFORNIA
MAY 1996

PAUL SWANSON, M.D., *was born in Minnesota in 1958. He graduated from Harvard University in 1980 with a major in biochemistry, the University of Minnesota Medical School in 1986, and a three-year residency program in emergency medicine at Bowman Gray Medical Center in 1989. Since 1989 Paul has been practicing emergency medicine, traveling, working with people in a variety of fields, and studying the human condition. He has a particular interest in human evolution and the true nature and care of human health.*

KARL J. ABRAMS

Preface

WHAT HAPPENED TO ME IN MAY 1995 CERTAINLY PIQUED my skepticism as a chemist and a college professor – yet there was no way I could deny the evidence of my body and spirit. Like so many of us, I had become resigned to the stress and fatigue of an overburdened professional and personal life, and like so many others I had sought remedy in health food stores.

The shelves that fairly groaned with the promise of vitamin vigor, nutritional nirvana, pure soul salves – I tried many combinations, but none could dent, for very long, the progressive malaise I felt, the fatigue and erratic depression that seemed the going price of living, working, striving. I looked forward to a long-awaited Sabbatical, but I knew that it would take more than time off from work to quicken my step and brighten my outlook.

So you can imagine my cynical amusement that May, when a good friend from Marin County, California, sent me what he described as his "miracle regimen for health" – a jar of greenish capsules that smelled of the sea. Hadn't I downed enough "miracle" formulas in the past few years to have tried them all?

And yet, trusting my good friend and knowing that he, too, was a searcher in the nutritional forest, I knew I had to try his regimen. And so I did, and in a matter of weeks I felt my weary cynicism give way to glowing belief and a new sense of well-being that I hadn't known since my childhood. If I can convince others that the secret to renewed, sustained health and vigor can be found in daily doses of a food substance so pure and so full of nutritional radiance, well, that's only because I'm a believer myself . . . yet more than a believer.

As a chemist and a man of science, I've been determined to unlock the scientific secrets of the superfood that changed my life, and I've made enough headway in that effort to share the truth with you. It makes sense, of course, in the elegant manner that only good science can explain, but all the same there's no disputing, and perhaps no questioning, the miracle of Earth's primary – and most potent – nutrient. Let me explain the chemical truths that are changing my life and the lives of other enlightened millions for the better – indeed, for the best.

Karl J. Abrams
MAY 1996

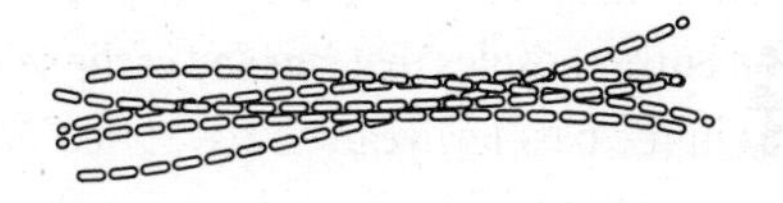

This book does not provide any medical diagnosis or treatment,

only dietary suggestions for good & vibrant health.

If you are ill, feeling ill, or on medication,

consult with a physician experienced in

the effects of dietary change and

other medical practices.

&

Table of Contents

Evolutionary Tree of Life

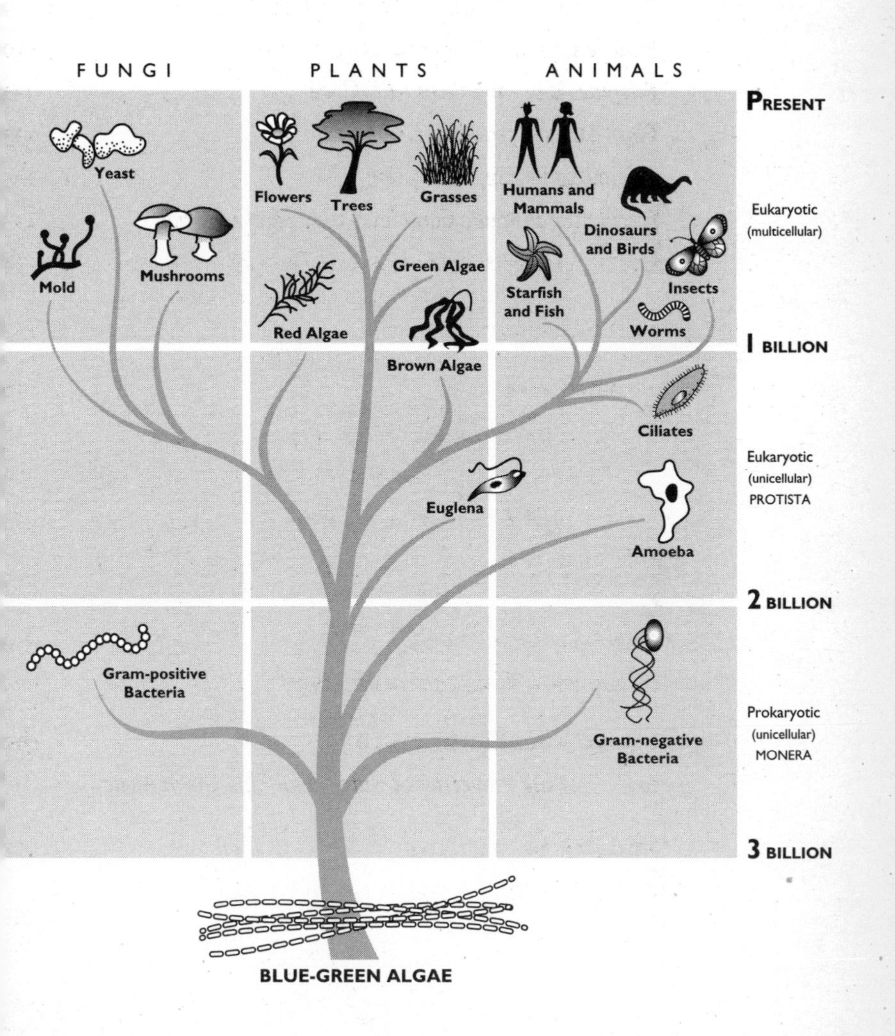

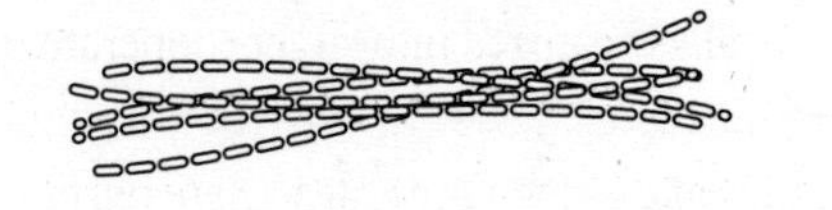

CHAPTER ONE

Algae – The Ancient Spark of Life

PRIMITIVE EARTH, STEAMY AND ENVELOPED IN DENSE GASES for billions of years, was covered with ancient oceans and pregnant with possibility for untold eons. Simple molecules such as amino acids, sugars, and fatty acids gradually and biochemically organized into larger cooperative molecular systems. Then, about three or four billion years ago – in a divine twinkle – magnificent, microscopic seeds of life called blue-green algae were born.

For the next billion years, oceanic algae species flourished, presumably as the only life form on earth. They flourished and inhabited wherever there was moisture, sun, and earth enough to sustain them. They changed and evolved; one variety turning the ocean into a "red tide"; another lit it up, making the new planet glow with mysterious phosphorescence. Slowly these algae invaded the land and enriched it, turning tropical soils green, thus preventing erosion. The algae began to produce a new gas called oxygen, and this set the stage for the evolution of all new life – thirty million different species of life – all that our sun, earth, and water could invent. And now, as always, the algae abundantly and continually sustain the beginning of our global food chain.

Today, by some estimates, there are more than 40,000 different species of algae. Ocean seaweed, such as kelp, can easily form colonies more than 100 feet long. Most of the water plants today, whether in fresh or salt water, are examples of algae species. There are algae virtually wherever there is water – iced or boiling – in the

The technical term APHANIZOMENON *(genus) and* FLOS-AQUAE *(species) are actually zoological terms of classification that literally mean "invisible living flower of the water" in Greek. To fully classify AFA blue-green algae scientifically is to state all biological categories from the largest to the smallest as follows:*

PHYLUM • *Cyanophyta*

CLASS • *Myxophyceae*

ORDER • *Nostocales*

FAMILY • *nostocaecese*

GENUS • *Aphanizomenon*

SPECIES • *flos-aquae*

desert, and even in the air. Some are found on glaciers and color snow red or green. There are even some – *Anabaeniolum* and *Simonsiella* – that inhabit the gastrointestinal tract of human beings and are considered part of our friendly microflora! They too are part of the family called blue-green algae, as are the species *aphanizomenon flos-aquae,* the heart of this book.

Primitive Earth – Preparing for Miracles

Algae, like the members of the plant kingdom that evolved from them, are able to combine CO_2 and H_2O, with the help of sunlight to synthesize glucose, the most important and ubiquitous of the simple sugars. They do this by use of their pigments, the most important among them being chlorophyll. The waste product of this process of carbon fixation is molecular oxygen, the gas that dramatically changed the atmosphere of Earth. From that point on, animals, including humans, were to evolve and live.

Billions of years ago, the atmosphere of primitive Earth was almost devoid of the oxygen gas we now take for granted. Although there were some friendly, nontoxic gases such as carbon dioxide and nitrogen, mostly there were hostile and poisonous gases such as ammonia, carbon monoxide, and hydrogen sulfide.

Fossils from rocks in central Australia, going back more than 3.5 billion years, tell us that early forms of algae were just beginning to slowly produce oxygen gas as a waste product. This gas was to be the life-sustaining gas of the entire animal kingdom that would evolve from the blue-green algae billions of years later.

Today we know, from rather interesting experiments carried out at the University of Chicago in the 1950s, that the energy available from ultraviolet light, volcanic activity, and lightning, along with the continual collisions among the minerals, gases, and organic fragments available on primitive Earth, was to create a wide variety of important biomolecules. These molecules would later serve as the basic building blocks of early blue-green algae and all life forms to follow. Some of those early biomolecules, as important today as they were on primitive Earth, are:

- *Sugar molecules, which chain together (polymerize) to form polysaccharides.*
- *Amino acid molecules, which polymerize to form polypeptides and larger proteins.*
- *Fatty acid molecules, which group together to form lipids (fats and oils).*

We discuss the nutritional biochemistry of these molecules in more detail in later chapters.

Forms of Algae

Brown Algae

Brown algae are the largest forms of algae, commonly known as seaweed or kelp. Found in cold ocean water, they are represented by more than a thousand species. Some float freely whereas others grow on the ocean floor. The free-floating variety use specialized air bladders to enable them to stay closer to the ocean surface to better access sunlight for their photosynthesis.

Rockweed flourishes on the rocks of coastal shorelines. This and other forms of brown algae (e.g., sargassum from the Sargasso Sea) are commercially used as a source of algin and alginic acid, which is used in ice cream, chocolate milk, toothpaste, and dripless paints. Brown algae are also a valuable source of elemental iodine. As a food source, brown algae are fed to cattle and sheep. The medical and technological uses for brown algae in general are quite amazing. Their use in time-release tablets and pill coatings, cough medicine, and disappearing surgical gauze just begins the list. Ashed algae have been used for centuries to help supply the missing minerals in soil or even animal feed.

Red Algae

All 2500 or so of the rhodophyta algae species use a special red pigment (phycoerythrin) to allow them to carry on photosynthesis deep within the sea, as deep as 600 feet near the equator. Rhodophyta (from the Greek *rhodos*, red) are a type of "seaweed" that are most abundant in warm tropical coastal waters. They have been harvested for centuries in Asia as a food source (Nori seaweed).

Green Algae

Algologists and botanists have classified about 7000 different species of green algae, mostly fresh-water varieties. They are all green because of their high concentration of chlorophyll. Green algae have been used to purify water and sewage systems because they can so easily remove toxic metals.

CHLORELLA are green algae that have been used as a food, as well as a source of antibiotics, in many parts of the world. They are unicellular spherical microorganisms with a variable protein content ranging from 15 to 60 percent.[1] Depending on the source, digestive problems with *Chlorella* are frequently reported because its outer cell wall is made up of nondigestible cellulose. All of the *Chlorella*-growing companies today process their algae so that the cell wall is disrupted.

PROTOCOCCUS are inedible unicellular green algae found as a slippery film on moist rocks or damp tree trunks. Spirogyra form the often-seen poisonous filamentous colonies of "pond scum" found in stagnant fresh-water ponds and lakes.

PLANKTON are made of green algae (desmids) and drift in vast colonies near the surface of ocean water. Those that photosynthesize are often called phytoplankton; those that do not are called zooplankton.

Blue-Green Algae, or Cyanobacteria

Blue-green algae are the simplest of the algae. They are best described as primitive and ancient bacteria with a characteristic green pigment (chlorophyll) that enables them to carry out photosynthesis. Because they lack a nuclear membrane, they are termed prokaryotic bacteria. *Oscillatoria* and *Spirulina* are well-known examples. On the other hand, eukaryotic cells, such as *Chlorella* and *Euglena* algae, are more complex and have a nuclear membrane.

Under a microscope, individual blue-green algae cells often look like a string of pearls surrounded by a gelatinous covering, which holds them together into filamentous colonies. Specialized cells within the colony are able to change the nitrogen of the air into the starting materials used to construct amino acids and proteins.

A study funded by the United Nations in 1993 described how blue-green algae are often collected and prepared by the women of Barkadrousso in the area of Lake Chad, Africa. Fighting off starvation and drought, these natives eat dihe, a protein- and mineral-rich meal of prokaryotic algae and chili peppers.[2]

According to paleobiologists such as J. William Schop, blue-green algae haven't really changed that much in the last billion years or so. In other words, blue-green algae are no longer evolving because they have, from an evolutionary standpoint, already achieved a kind of biological perfection.[3]

The green chlorophyll pigments that characterize plants today theoretically originated from ancient tiny blue-green algae cells that slowly, over time, became incorporated into larger cells and now serve them as organs of photosynthesis.

Aphanizomenon flos-aquae (AFA): Our Super Hero

One particular species of blue-green algae, *Aphanizomenon flos-aquae* (pronounced *"A-fan-is-zóh-me-non floss ah-kéw-ay"*) or simply AFA, flourishes today in the beautiful waters of Upper Klamath Lake in southern Oregon. AFA algae probably began to grow and bloom in this lake around 7,000 years ago, just after the time when retreating glaciers carved out a shallow lake fed by seventeen pure and pristine rivers and streams, as well as the very clear and pure underground water system fed by nearby Crater Lake. The pure water availability of Crater Lake – the deepest (3,000 feet!) yet clearest lake in North America – and the vast mineral contents of Upper Klamath Lake were brought together and made available to AFA algae through a cataclysmic volcanic eruption at about the same time as the lake's formation. Nearby Mount Mazama erupted with a force estimated by geologists to be hundreds of times greater than any eruption of the twentieth century. The mineral content deep within the earth surfaced explosively through the air for hundreds of miles, providing the topsoil of the Cascade Mountains with incredibly rich and diverse minerals.

The yearly rains then washed these minerals directly into the very wide area of Upper Klamath Lake, building a nutrient-dense and muddy lake bed sediment ranging between 30 and 40 feet deep. With more than 300 days per year of photosynthesizing sunlight, AFA flourished in the alkaline lake and bloomed four times yearly as nowhere else on Earth. In this lake we have what is probably the last of Earth's pure and wild superfoods, and enough of it to substantially help nourish the inhabitants of the Western Hemisphere.

In a *Scientific American* article, algologist and professor of aquatic biology at Wright State University Wayne Carmichael discusses the nature of AFA cells and their connection to the acidity and alkalinity of the waters they inhabit. The high alkalinity of Upper Klamath Lake (pH = 9 to 11) insures the health of our AFA algae.[4]

Few people are probably aware of the fact that the mountains surrounding Upper Klamath Lake contain one of the largest areas of old growth forests in North America.[5] Within the lake and surrounded by this ancient mineral-rich paradise, AFA algae have flourished for millennia.

According to Carmichael,

> *"Cyanobacteria are known, too, for the critical insights they have provided into the origins of life and into the origins of organelles (cellular organs) in the cells of higher organisms.... Because they were the first organisms to carry out oxygenic photosynthesis, and thus to convert carbon dioxide into oxygen, they undoubtedly played a major part in the oxygenation of the air."*

AFA – The Heartiest Food on Earth

Why is AFA algae so powerful a "super-food" and so effective in transforming the health status of humans to vitality and boosted immunity? It is partly because of the conditions under which AFA flourish, free from the technological pollutants of insecticides, pesticides, herbicides, and chlorinated phenyls and dioxins.

It is also because cyanobacteria prokaryotes such as AFA are "heartier" than red, brown, or green algae, and seem to better tolerate environmental extremes. But why is AFA so able to withstand such extremes? After evolving so successfully for eons, AFA has developed wide and versatile enzyme systems. When we eat AFA algae we benefit accordingly. A variety of enzymes means a generous selection of their constituent minerals, vitamins, and amino acids, all of which are free from dangerous environmental pollutants.

There are other naturally protective substances in AFA. The glutathione molecule, made from three interlocking amino acids, was probably Earth's first antioxidant. It was used to shield and protect primitive forms of blue-green algae from the burning effects of a harsh and hostile environment. For humans, the same glutathione from AFA has been shown to protect us from the toxic effects of a polluted environment.

Free Radicals – Beware the Fire Within

As primitive Earth slowly transformed into a planet with an oxygen-containing atmosphere, sugars, lipids, and amino acids could be – for the first time – efficiently oxidized and harnessed for their life-giving energy. Ironically, along with the availability of atmospheric oxygen to give life also came several dangerous forms of oxygen that could take life away. These were the highly reactive cousins of oxygen – now called free radicals – which did not have all of their electrons paired up as they should. Thus, they had an odd number of electrons. As a result, these "unstable" free radicals were able to crash into delicate biomolecules within cells and rip off the single electron they were so desperately missing. This process of stealing electrons from cellular biomolecules repeats again and again, creating still more free radicals, as in a chain reaction. Physical chemists report that the impact of these free radicals careening into chromosomes, cell membranes, and other important parts of a cell actually produces tiny bursts of light as crucial bonds are destroyed.[6]

Early cellular life, therefore, to insure and protect its very existence, had to evolve a family of specialized molecules called "antioxidants" that would be able to act as protective shields and stop the continuous damage of free radicals. Blue-green algae succeeded by evolving antioxidant shields such as betacarotene. Other organisms that did not, eventually perished. Today we can derive this same antioxidant protection by ingesting organisms that ingeniously survived the free radicals to this day – the blue-green algae.

Modern Diseases and Natural Cures

Many of our prescription drugs today were either isolated from plants at one time or carefully created in the laboratory to mimic their healing properties. We should not be too surprised to learn that the biostimulating and health-enhancing properties of algae's "nutritive fine chemicals" are being carefully studied today all over the world.[7] After all, land plants and the medicines derived from them have ultimately evolved from their ancestral blue-green algae.

Even the National Cancer Society is looking into the "immunomodulating" properties of AFA and other blue-green algae as a way to combat cancer. In March

1996, the powerful pharmaceutical laboratories of Eli Lilly and Co. announced ambitious plans to try to patent any and all isolatable anticancer compounds that can be found in certain species of cyanobacteria. They were not willing to divulge which particular species held the most interest.

Sometimes we can be nutritionally rescued by common and simple plant extracts. Raw cabbage juice, for example, has had remarkable success in treating ulcers.[8] This is probably because of the high glutamine (an amino acid) content of the cabbage. It is thought that this amino acid stimulates the production of proteins, which actually protect the stomach lining. Of the twenty amino acids in AFA, glutamine content is the highest at 76 milligrams per gram of flash-frozen, freeze-dried AFA algae. Perhaps this may substantiate many of the anecdotal reports that the algae serves well to protect us from stomach ulcers.

By some estimates, about half of the U.S. population is suffering from food and environmental allergies that are fast becoming the leading cause of undiagnosed symptoms. Such allergies put considerable stress on the immune system. They often develop from eating too much of the same foods day after day, causing low immune function in the gastrointestinal tract, which is then unable to stop the intrusion of foreign microbial invaders. In fact, chronic fatigue syndrome may be caused by the Epstein-Barr virus becoming activated by a highly stressed and weakened immune system.[9]

We must keep our immune system strong for many reasons. A strong immune system may help us to resist being infected by these and many other modern afflictions. If already infected, a healthy immune system may help guarantee that a variety of microbes remain dormant and harmless because they never have a chance to begin their cycle of destruction. AFA offers us such protection. The remaining chapters of the book will show why we call this book *Algae to the Rescue!*

Environmental Pollution and Heavy Metal Problems

Heavy metals such as mercury, aluminum, lead, and cadmium are toxic to the kidney and other vital organs. Amino acids present in the algae and bioflavonoids (such as anthraquinones) are very good at binding heavy metals to them in the urinary tract. Some amino acids in AFA algae are able to help remove from our system such heavy metals as toxic lead and cadmium ions, which otherwise would build up in our bloodstream.[10] Heavy metals such as these seem to have a strong role in learning disabilities in children and may even contribute to increases in criminal behavior.

There are thousands of ways in which a chemically polluted environment can impact the status of our general health. There are probably about 5000 food additives used in the United States, many of which have been linked to learning disabilities and behavior disorders, especially in children.[11] Other countries, for example Canada and Australia, have severely and rightfully restricted the use of such additives. Progress has been too slow in the United States, however, and for frustrating reasons. Our very powerful "Nutrition Foundation" claims that there is not yet enough data to cause concern. The U.S. public probably does not as yet know that much of the funding and support for this foundation comes from the manufacturers of most of these very same food additives.

The average person consumes enough sulfite additive every day to induce asthma, hives, and other common allergic reactions.[12] Sulfite is ubiquitous because it is typically sprayed on fruits and vegetables and added to most processed foods to prevent spoilage. Oxidizing agents such as sulfites can also increase our allergic reactions to still other agents, thus compounding the problem.[13]

In 1996 The World Resources Institute, an environmental research group in Washington, DC, reported that when humans are exposed to environmental pollutants there is also a dramatic reduction in the strength of their immune systems. This institute has asked the World Health Organization to look into the connection between this global public health threat and the immunosuppresion effects that weaken the body's resistance to infection and cancer.

Pigments in AFA are colored molecules, vital to AFA's photosynthetic powers.
Each of its four colored compounds is responsible for transforming
a different wavelength of the sun's light, and each color
has its own unique effect on human cells.
The orange carotenes – the beta type being the most famous –
biochemically inspire intestinal and lung cells to regenerate themselves.
Green cholorophyll – as well as the red & blue pigments that
algae alone offers – protects our cells from the
fiery dangers of colliding free radicals.

CHAPTER TWO

Aphanizomenon Flos-Aquae: A Chemist's Look Inside

IF WE REALLY WANT TO FULLY APPRECIATE THE NUTRITIONAL benefits of AFA algae, we need to understand a little of its chemistry and what happens when humans ingest it. The effects of the nutritional ingredients in AFA appear to be profoundly and synergistically interdependent. For example, without the small amounts of vitamin E present in AFA, vitamin A levels in the blood would always remain low no matter how much supplementation is ingested. With too much zinc, not enough copper would be absorbed. With a little more or a little less of any one of the twenty amino acids already in AFA, amino acids would no longer be delivered to our nervous systems in just the right mood-elevating proportions.

Basic Biochemistry Made Simple

To begin with, AFA algae is 95 percent water, H_2O, before it is flash-frozen and freeze-dried. The chemical formula for water tells us there are two atoms of hydrogen electrically bonded to a single, central atom of oxygen. The resulting cluster of three atoms is called a molecule of water. Its molecular structure is usually depicted by "lines" that symbolize the chemical bonds that hold two atoms to each other:

$$H\text{-}O\text{-}H$$

Many of the water molecules inside the AFA cell are "hydrated" to mineral ions such as potassium, sodium, or calcium in that they are surrounded by water on all sides. Along with hydrated mineral ions, there is a wide assortment of organic molecules ranging in size, shape, and function.

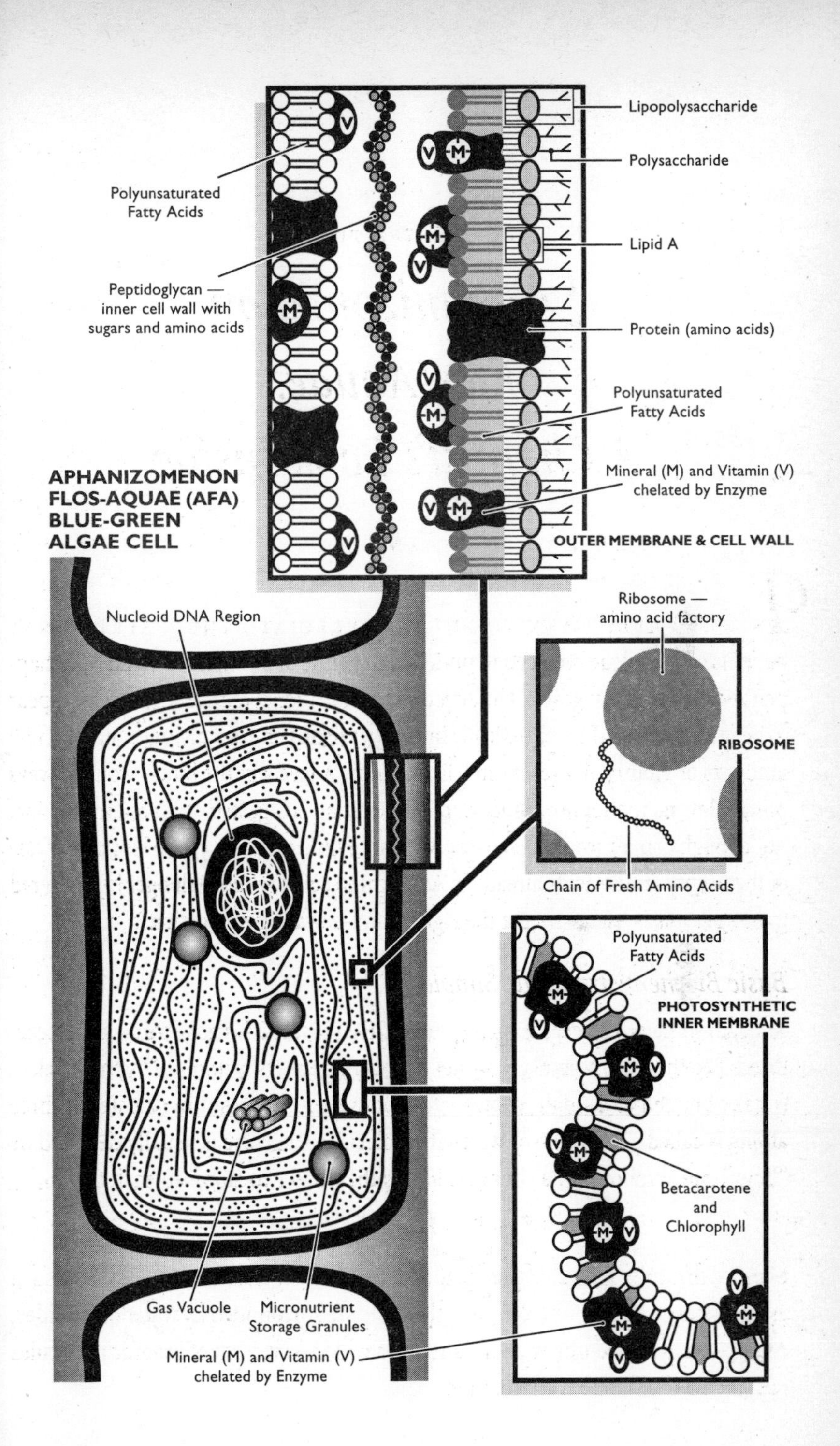
APHANIZOMENON
FLOS-AQUAE (AFA)
BLUE-GREEN
ALGAE CELL
Polyunsaturated
Fatty Acids
Peptidoglycan —
inner cell wall with
sugars and amino acids
Lipopolysaccharide
Polysaccharide
Lipid A
Protein (amino acids)
Polyunsaturated
Fatty Acids
Mineral (M) and Vitamin (V)
chelated by Enzyme
OUTER MEMBRANE & CELL WALL
Nucleoid DNA Region
Ribosome —
amino acid factory
RIBOSOME
Chain of Fresh Amino Acids
Polyunsaturated
Fatty Acids
PHOTOSYNTHETIC
INNER MEMBRANE
Betacarotene
and
Chlorophyll
Gas Vacuole
Micronutrient
Storage Granules
Mineral (M) and Vitamin (V)
chelated by Enzyme

The simplest part of these organic molecules is their hydrocarbon portion. Hydrocarbons consist of a zigzag chain of interconnecting carbon atoms, each of which is bonded to enough hydrogen atoms to give a total of four bonds. The hydrocarbon portion, in turn, is often attached to still other more important clusters of atoms called functional groups. These groups impart special properties to all hydrocarbon molecules. Proteins, amino acids, carbohydrates, sugars, and fatty acids are well known examples of hydrocarbon molecules attached to one or more functional groups. Probably, the three most important functional groups in all biochemistry are:

- *Acid groups (fatty acids, amino acids)*
- *Alcohol groups (sugars, carbohydrates)*
- *Amino groups (proteins, enzymes, amino acids)*

By attaching an acid group to a hydrocarbon chain, a fatty acid is created; by attaching one alcohol group to a hydrocarbon chain, various alcohols are created; by attaching many alcohol groups, sugars are created; by attaching an amino group *and* an acid group to a hydrocarbon, an amino acid is created.

Carbohydrates are really nothing more than a long chain or polymer of bonded sugar molecules; proteins, similarly, are just a long chain of amino acids.

Cell Membranes – The Source of Flexibility

About four billion years ago, in the ancient oceans of Earth, a wide assortment of simple, yet biochemically important fatty acids were being slowly formed into primitive cell membranes necessary for the protection and maintenance of life and the eventual formation of blue-green algae. The cell membrane slowly evolved into a semi-fluid, multilayered covering, which today surrounds the body or cytoplasm of every cell, including its many smaller internal bodies, called organelles.

These membranes typically consist of two molecular layers of lipid molecules, each of which needs to be "fluid and flexible" if the cell itself is to remain healthy. Lipids – also called fats – along with carbohydrates and proteins, are required by all humans for proper nutrition. Most of the foods we eat contain some form of lipid. Meats and dairy foods contain one form, and nuts, seeds, and vegetables have another. Some lipids can be synthesized by humans; those that cannot are termed essential fatty acids (EFAs).

Each lipid molecule is made up of two regions. One is a densely packed, somewhat spherical region consisting mostly of oxygen, phosphorous, and sometimes nitrogen. This results in a small and spherical water-soluble "head." Because there is phosphorous in the head of the membrane's lipid, the entire lipid is sometimes called a phospholipid. Phospholipids are found repeating across all cell membranes, the spherical heads of which form a kind of cobblestone protective covering around the cell. The second and lower region of the lipid consists of two hydrocarbon chains of variable length, forming two long, water-insoluble "tails."

The outer and inner layers of the cell membrane also contain various kinds of proteins, all of which are flexibly imbedded in the membrane and nearly completely cover the lipids beneath. Some of the specialized proteins of the cell membrane actually "poke through" both sides of the membrane and form cylindrical pores that allow mineral ions such as sodium and potassium to pass through. When ions are chemically pumped across a cell membrane, either out of or into the cell, a measurable ionic current is actually created. This current is especially vital to the proper activity of brain and muscle cells.

The fatty acids that form the lipid layers of the cell membrane ultimately affect the membrane's fluidity, depending upon the actual shape and length of each fatty acid. Fatty acids that are saturated – meaning each carbon atom has the maximum number of bonded hydrogen atoms – tend to be somewhat straight and lead to cell membrane molecules that are too rigidly packed together, which in turn leads to less healthy cells.

The sixty-four or so micronutrients in AFA are like the musical instruments of a biochemical orchestral symphony. When playing together the full power and beauty of each individual nutrient instrument is optimally realized. The first violinist and solo pianist might be the betacarotene and vitamin B_{12}, another entire section of the orchestra might be the synergistic effect of the amino acid or vitamin B family. What factors determine which music is to be played and by which conductor? AFA has three or four billion years of melodious and rhythmically evolving cellular memory to share with humanity. Could the choir be humanity and the conductor of this orchestra be evolution itself?

On the other hand, fatty acids that are unsaturated are those that have one or more of their carbons bonded to less than the maximum number of hydrogen atoms. Their resulting lipid molecules cannot be rigidly packed together. This leads

AFA is a living periodic chart of synergistic bioelements that no human chemical laboratory could ever synthetically duplicate.

to a favorable condition – a fluid and flexible cell membrane. Such cell membranes give us healthy skin and brain cells. Diets that include AFA blue-green algae introduce a high proportion of unsaturated fatty acids, which help to maintain the fluidity and health of our cell membranes.

Betacarotene Strengthens Cell Membranes

Betacarotene is known to inhibit viruses by strengthening the membranes of cells that may be under attack.[1] Other similarly useful carotenoid compounds (such as lutein, zeaxanthin, and alpha-carotene) have also shown equally potent antiviral as well as possible antitumor effects.[2] There is mounting evidence that all carotene compounds also have protective properties that help to protect and thus maintain the strength and resilience of the cell membranes that comprise the epithelial cells of our respiratory tract.[3] AFA is rich in carotenoid compounds.

Storage Granules – Morsels for the Cell

If we look more closely inside an AFA algae cell under the powerful gaze of an electron microscope (thousands of times more magnifying than any ordinary light microscope), we see varieties of numerous, tiny storage granules spread about the cytoplasm:

- *Cyanophycin storage granules are made of polymers (repeating chains) of just two amino acids: aspartic acid (asp) and arginine (arg). These plentiful granules serve as a polypeptide energy reserve when other nutrients are unavailable. They can be as high as 20 percent of the dry weight of the AFA cell. The molecular structure of the cyanophycin polymer tells us that each aspartic acid molecule is bonded into a long chain, each of which is bonded to a single molecule of arginine.*

- *Both amino acids in the storage granule of AFA work together to biostimulate the thymus gland and enhance our ability to resist infection. Such biostimulation may even cause this gland to increase in size, even though it tends to shrink dramatically as we grow older. As a result, there is an increased production of healthy white blood cells. This greatly boosts the immune system and can be helpful in the treatment of chronic fatigue syndrome or the prevention of cancer. The arginine amino acid also biostimulates the pituitary gland to produce hormones that signal the liver to synthesize detoxifying and liver-protecting proteins.*
- *Glycogen granules are made of polymers of glucose, a monosaccharide sugar. This polysaccharide serves as a carbohydrate energy reserve.*
- *Polyphosphate granules are made of polymer chains of phosphorous and oxygen. For AFA algae (and humans), phosphorous is essential for the construction of membranous material that is critical to the health and maintenance of each cell.*
- *Cylindrical granules are composed of dozens of stacked discs called thylakoids. Each disc is actually a hollow, flattened sac, the wall of which is made of a membranous material that contains the molecular machinery required for the miracle of photosynthesis.*

Attached to the thylakoid membranes are granule-like clusters of two phycobiliprotein pigments, which constitute about 10 percent of the cellular mass.[4] One of them gives a reddish coloration (phycoerythrin) to algae, the other a bluish coloration (phycocyanin). The "phyco" part of the name actually refers to algae; "erythrin" refers to the color red and "cyanin" indicates the color blue. Both pigments are believed to be nutritionally beneficial phytochemicals, probably because of their antioxidant and free radical quenching abilities. Such phytochemicals are currently the focus of exciting nutritional research.

Photosynthesis – Algae's Innovation with Light

All life in Earth's biosphere ultimately depends upon the biochemical assimilation of sunlight in the form of quantum energy packets called photons. Plants and algae transform carbon dioxide and water into sweet high-energy glucose and our life-giving oxygen gas. This transformation is carried out by a complex chain of chemical events collectively known as photosynthesis.

The molecular pigments involved in photosynthesis include green chlorophyll and proteins such as blue phycocyanin, which arrange themselves as colored particles on the inside surface of the AFA cell membrane. This, of course, explains the blue-green color of AFA algae. Other algae proteins, however, are red or brown, and other carotenoid pigments, such as betacarotene, are yellow or orange. Varying amounts of these pigments yield a wide range of colorful microalgae. The Red Sea gets its color (and its name) from such microalgae.

But where does the O_2 we breathe come from? Does it originate from the oxygen atoms in carbon dioxide or does it come from the oxygen atoms in water? Clever scientific studies demonstrate that the O_2 we breathe was originally inside the H_2O molecule itself. This means that photosynthetic chlorophyll is able to strip water of its hydrogen atom and connect it to CO_2, turning the carbon, oxygen, and hydrogen atoms into molecules of glucose for food. Amazing!

Actually about 80 percent of all photosynthesis on this planet is carried out by one or another algae species. Half of the photosynthesis in the oceans and lakes of the world is carried out by blue-green algae, our earliest pioneer of this life-giving and life-sharing process.

AFA algae still carries on its photosynthesis within ancient yet efficient bio-factories called thylakoid sacs. Their photosynthetic efficiency is reflected by the fact that no other life form has such generous amounts of the transformative pigments, carotene and chlorophyll.

To understand the events of photosynthesis in AFA, we must look deeper within the cytoplasm, inside the thylakoids. These disc-like sacs lie one on another to form cylindrical structures (called granum) into which CO_2 and H_2O molecules enter, and from which sugar molecules emerge.

With the help of an electron microscope, we begin to see tiny dots scattered over the outer membranes of the thylakoid sacs. Zooming in still closer, these dots become complex factories which contain stacks of chlorophyll molecules that can assimilate and feed upon light. Other pigment molecules, such as the betacarotenes and phycobilins, also absorb light, and, by doing so, help protect the AFA algae from the damaging effects of too much sunlight.

Hundreds of chlorophyll molecules are neatly stacked together with dozens of betacarotene molecules. Collectively they act as one light-sensitive bioantenna which responds to the impact of incoming sunlight. Chlorophyll and betacarotene collectively are among several antioxidant protective factors that inhibit mutation-promoting effects of a variety of carcinogens. Both substances protect the DNA of algae and humans alike during the very vulnerable time of cell division.

Millions of such chlorophyll-betacarotene bioantennae are snugly imbedded within the membrane of the thylakoid discs. Each bioantenna, upon absorbing a single photon of reddish light from the sun, is then able to eject a single electron from one of its chlorophyll molecules. That negatively charged electron is then captured and passed from one receptor molecule (pheophytin) to another (plastoquinone) until the latter attracts an oppositely charged positive hydrogen from the fluids inside the cell. This decreases the cell's acidity.

The chlorophyll molecule, with the help of the amino acid tyrosine, now recovers its lost electron by removing one from a nearby mineral source – the manganese atom. That same manganese atom then "breaks open" a nearby water molecule and regains its lost electron. This causes the H_2O to split into oxygen gas and positive hydrogen ions, which build up inside the thylakoids. This increases the cell's acidity. Finally, hydrogen ions migrate from the inner region of high acidity to the outer region of low acidity. Just as a harnessed waterfall in a hydroelectric dam produces electrical energy, the hydrogen ion "waterfall" supplies the energy needed to synthesize glucose as food.

Carbohydrates – Source of Energy

AFA algae is about 25 percent carbohydrate, most of which is found in the cellular covering called the cell wall. Carbohydrates are manufactured by photosynthesis and secreted by the cell for environmental protection. There are three major categories for all carbohydrates:

- *Monosaccharides are simple sugars that are composed of single molecules (monomers). The word "saccharide" is Greek for sugar. The most abundant monosaccharide is glucose. Galactose, mannose, and ribose sugars are other monosaccharides found in AFA. (The names of all sugars typically end with -ose.)*
- *Oligosaccharides are sugars that consist of short chains of two to nine monosaccharides. Lactose, or milk sugar, is a typical disaccharide consisting of two monosaccharides, glucose and galactose.*
- *Polysaccharides consist of long chains of sugars ranging from ten to thousands of monosaccharide units. Most carbohydrates found in nature occur in such long chains. The indigestible cellulose of most plant cell walls that is found in wood, paper, and cotton is a linear polysaccharide of glucose. Starch and glycogen are branched polysaccharides of glucose. Polysaccharides such as starch (and glycogen) are actually glucose polymers used by plants (and animals) to store food.*

Cellulose is a polysaccharide used to form rigid cell walls. Each wall contains thousands of glucose molecules bonded so strongly as to be indigestible. Simple carbohydrates such as glucose are considered to be food because they can be biologically oxidized in our body to produce H_2O, CO_2, and energy. The word carbohydrate originates from the fact that when these sugars were heated in a test tube, a residue of black carbon and water (hydrate) remained.

Inner (Peptidoglycan) Cell Wall – Interlocking Genius

The cell walls of plants and algae alike are made of a complex material that is synthesized and secreted by the cell to protect its delicate and flexible cell membrane. The cell wall strengthens and protects the cell, but it is also flexible enough to withstand stress and allow for cellular growth. In general, the cell walls of all plants and algae are remarkably similar both chemically and physically. This is because higher plants were derived from the simpler blue-green algae many millions of years ago.

The cell walls of plants and some microalgae such as chlorella are composed of overlapping polymers of 10,000 or so tightly bonded simple glucose molecules, collectively called cellulose. Since humans do not have the enzymes necessary to digest cellulose, the individual glucose molecules cannot be used as a source of energy.

In contrast, the basic structure of AFA's cell wall is that of a thin, soft, and very digestible polymer. This polymer is a *peptidoglycan* material made from polypeptide chains of four amino acids that hold together the longer polysaccharide chains. This glycan portion consists of thousands of alternating sugar-like molecules that are very similar to glucose and galactose. The actual shape of the AFA cell is determined by the beautiful geometric arrangement of these interlocking chains.

Outer Cell Wall – A Cellular Barrier Reef

The outer layer of the cell wall is held together by a less understood cross-linking of polysaccharides,lipids, and proteins. This portion of the cell wall serves as an outer barrier that allows only small molecules such as H_2O to pass through, yet prevents large and valuable enzyme molecules from escaping. Some of the immunostimulating properties of AFA have been traced to the complex lipids of its outer wall.This subject is treated in more detail in *Appendix A*.

Chlorophyll and Hemoglobin – The Algae Nexus

CHLOROPHYLLS are complex green pigment molecules that can absorb photons of electromagnetic energy (light). One of these molecules, chlorophyll, is found in higher plants as well as AFA and other blue-green algae. Its chemical formula, $C_{55}H_{72}MgN_4O_5$, reflects the size and complexity of the molecule. Its single and lone magnesium ion sits snugly in the center of a flatter region of a beautiful, ring-shaped molecule called porphyrin. Dangling from this ring is a long and fat-soluble hydrocarbon chain, which is used to connect the entire chlorophyll molecule to specialized cell membranes (the thylakoids) deep inside the cell. The entire structure acts as a light-absorbing "antenna" and plays a crucial role in the miracle of photosynthesis.

HEMOGLOBIN is an even more complex molecule, which can transport oxygen molecules from the air to the cells within higher animals. Its chemical formula – $C_{34}H_{30}N_4Fe$ – reflects the fact that it is essentially a protein molecule connected to a chlorophyll-like porphyrin ring. Instead of a magnesium ion, however, there is a snugly held iron ion at the center.

It is apparent that chlorophyll and hemoglobin are remarkably similar in structure, although widely different in function. While it is true that both molecules are vital enzymatic pigments, it is not yet apparent how ingesting chlorophyll can somehow enhance the blood oxygenating function of hemoglobin. Still, there may well be a connection. When the porphyrin ring of either chlorophyll or hemoglobin is broken, straightened out, and connected to a protein molecule, what is left is a structure called phycobilin. It is this pigment that imparts to cyanobacteria

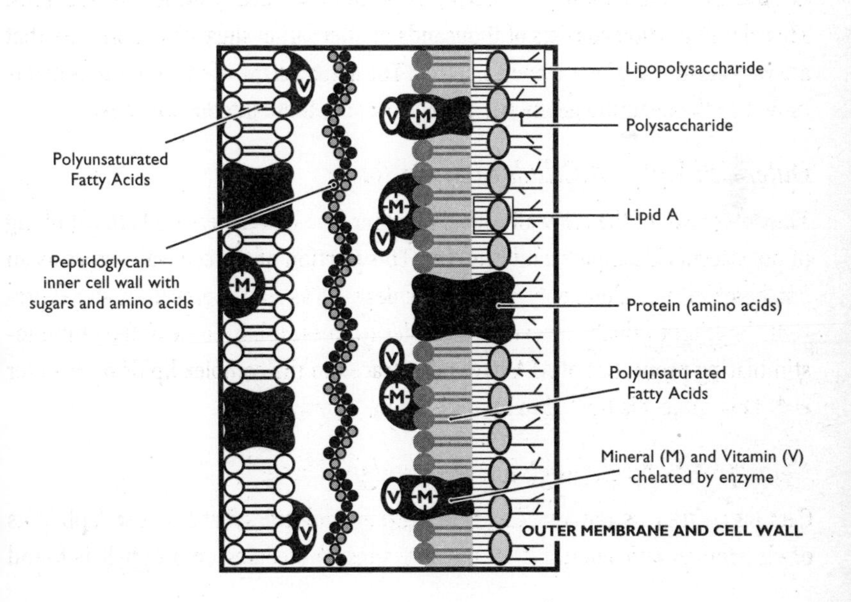

AFA algae (cyanobacteria) provides our only opportunity to digest the vast nutritional cornucopia in this profoundly unique and miraculously tiny cell without a nucleus. By containing chlorophyll, like a plant cell, and by having motility like an animal cell, AFA manifests all kingdoms of the biosphere at once!

their well-known shades of blue and red, depending on the shape and concentration of the connected protein.

It is true that many people have reported feeling refreshed by chlorophyll. Some claim that the 2.5 percent chlorophyll in AFA algae is more than enough to provide the raw materials needed to help oxygenate the blood. There have been some healing-related experiments done with this green molecule of AFA. Heavy menstrual periods, for example, have been successfully treated by chlorophyll supplements using only 25 mg per day.[5] This would correspond to the chlorophyll content of roughly one daily gram of AFA.

Part of the magic of chlorophyll comes from the relatively rare magnesium it contains. Being part of approximately 325 different human enzyme systems truly makes magnesium one of the greatest enzyme bioactivators. Good and verifiable research needs to be done. Presently there is work being done in Japan to determine if chlorophyll and hemoglobin molecules can be "inter-biosynthetic."

DNA and RNA – Algae Nucleic Acids for the Mind

Thanks to a single strand of the double helix deoxyribonucleic acid, known as DNA, genetic information may be transcribed and transferred to the smaller, but equally amazing ribonucleic acid, known as RNA. Provided all communication lines are working properly and there is little free radical damage to the DNA and RNA, AFA's 20 amino acids can be correctly lined up and connected like pearls on a living neck-ace or letters of a biochemical alphabet. By this means, all the proteins and nzymes needed for the life of a healthy organism can be created.

As we grow older we are increasingly subjected to the cumulative effects and lestructive powers of random free radical damage to our DNA and RNA. The high oncentration of nucleic acids in AFA algae continually provides the needed "spare arts" to replace worn-out portions of these essential nucleic acids. As a result, ur entire physiology is bioregenerated.

In laboratory tests, animals injected with RNA / DNA lived almost twice as long s uninjected animals.[6] Even mice with tumors lived longer. More research needs o be done, however, to determine what percent of the ingested nucleic acids are urviving the journey through the gastrointestinal tract and actually getting bsorbed through the intestinal wall. Recent studies have shown that nucleic acid nolecules do somehow squeeze through. According to a 1995 study by the National Academy of Sciences, some portions (nucleic acids) of RNA / DNA nutrients are lirectly absorbed. They thus serve as "conditionally essential organic nutrients" hat can be reassembled back into RNA or DNA.[7]

It should be no surprise that the nucleic acids of AFA algae also biostimulate the mmune system. In fact, animals fed enriched diets of nucleic nutrients had substantially lower susceptibility to bacterial infections than their untreated counterparts.[8]

THE AFA CELL IS TRULY AWE INSPIRING. Although *hundreds of imes* smaller than the cells of the plant or animal kingdoms, its wide range of nicronutrients is remarkable and unsurpassed by any known food. From the hemist's point of view, the cell wall and inner cell membranes are richly composed of essential fatty acids and immunostimulating compounds. Packed inside the AFA cell are storage granules containing a balanced array of powerful protective antioxidants along with light-transforming pigments which support the cell as its ribosome factories ceaselessly manufacture its elegant proteins and enzymes. In the next chapter we explore these proteins and enzymes, and the amino acids that compose them.

CHAPTER THREE

Earth's First Protein

PROTEINS ARE THE MOST ABUNDANT TYPE OF BIOMOLECULE I any living cell, typically making up about half or more of any cell's dry weight. Hun dreds of different proteins are found in any one cell, and many thousands of dif ferent proteins carry on just as many different functions in larger organisms suc as humans.

Elegantly enough, all proteins, whether in AFA algae or in humans, are poly mers constructed from the same basic set of 20 amino acid monomers. Shor polymers of less than 60 amino acids are termed polypetides, and longer one are called proteins or enzymes. It helps to think of the 20 amino acids as let ters of a biochemical alphabet, and the polypeptides, proteins, and enzyme as the words they form.

Proteins – Pantheon of Amino Acids

There are about 100,000 different proteins in humans and only several thousan in AFA. Although fewer in number, those in AFA are easier to digest than those c higher plants or animals because they are smaller and less complex. Even so, AF algae still probably has the highest protein content of any known food. Some pro teins are found in AFA's cyanophycin storage granules and serve as food durin times of stress. Most are found embedded within the cell membrane. Other "trans verse" proteins completely penetrate the membrane and have pores that allow min eral ions and small molecules to pass through the cell. Altogether there may b several thousand different proteins carrying out one or another function.

Once ingested, AFA's proteins are small enough to be easily broken down int smaller, absorbable peptides and amino acids. They may then be used to help construct one or another of the 100,000 or so different kinds of human proteins.

Some proteins are long and fibrous and used as collagens, elastins, and keratins; other proteins, such as hemoglobin or transferrin, are spherical and used for transport in the blood. The rich and diverse protein content of AFA algae greatly helps to stabilize our blood glucose levels and supply amino acids, which build and repair our bodily tissues.

Amino Acids of AFA – The Healing Units Within

Along with phospholipids, amino acids are the building blocks of AFA's elaborate and flexible cell membrane system. Amino acids are found in all life forms, and are the most plentiful substance of living organisms, second only to water. Amino acids have both acid and amino functional groups that, when bonded together, create a diverse variety of alga polypeptides, proteins, and enzymes. All contribute to important and life-sustaining cellular functions.

Approximately 65 to 70 percent of the dry weight of AFA's protein can be attributed to twenty different amino acid molecules. No food on Earth has so much in a form so easy to assimilate. Apparently, not all amino acids in plants or animal tissue are equally bioavailable. Some grains are difficult to completely digest. Their proteins are coated with so much cellulose that human enzymes cannot easily penetrate them. Animal protein is difficult to digest because of its large size and molecular weight. Only free amino acids or small proteins are easily absorbed across the intestinal wall. The protein in AFA algae is actually the best of both groups. Their small size, low molecular weight, and nonassociation with cellulose makes AFA's amino acids ideal for ease of digestion and assimilation.

> *When any polypeptide enters a human cell, it is digested (technically hydrolyzed) by a specific enzyme (peptidase) down to its component amino acids. Strangely enough, the larger polypeptide sometimes enters the cell more readily than the more isolated amino acid. Thus amino acid supplementation should come from a natural source in which the amino acids are bonded into low molecular weight polypeptides, as in prokaryotic cyanobacteria.[1] People who have digestive difficulties, possibly from an enzyme deficiency, may not experience an optimum "profiled assimilation" of proteins. Chapter 14 discusses this problem further.*

AFA algae can easily biosynthesize all twenty amino acids, whereas higher plants typically lack one or two, and humans may lack as many as ten. Amino acids that humans cannot biosynthesize are called *essential*, and those that humans can biosynthesize are termed *nonessential*. See table below for a list of the amino acids available from 2.0 grams of flash-frozen, freeze-dried AFA.

Amino Acid Content of AFA Algae

Amino Acid (Essential)	*weight (mg)*	*Amino Acid (Nonessential)*	*weight (mg)*
Arginine	76	Asparagine	95
Histidine	19	Alanine	93
Isoleucine	59	Glutamine	152
Leucine	104	Cystine	4
Lysine	69	Glycine	59
Methionine	15	Proline	57
Phenylalanine	51	Serine	59
Threonine	65	Tyrosine	35
Tryptophan	15	Aspartic Acid	15
Valine	64	Glutamic Acid	8

It is interesting that AFA algae is able to biosynthesize all twenty amino acids in just the right proportions to maintain the health and fluidity of its cell membrane and its cytoplasm. But what is most amazing is that, according to the National Academy of Sciences, the composition or amino acid "profile" of the AFA is nearly exactly in the same proportions required by humans.[2]

In general, amino acids are sometimes biochemically modified by removing either the amine or the acid functional groups. When the amine is removed, the remaining molecule is used as fuel, when the acid portion is removed, the molecule may serve as a messenger (hormone) molecule. For example, AFA's tryptophan may be modified to serotonin, a brain hormone or general messenger molecule, partly through the assistance of an enzyme that removes tryptophan's acid group. Serotonin has a variety of effects, such as blood vessel constriction, intestinal peristalsis promotion, and neurotransmitter capabilities for certain types of nerve cells. Serotonin also increases the permeability of sugar across the cell membrane. Tryptophan and phenylalanine are two essential amino acids that can be biochemically rearranged by humans to create neurotransmitters associated with a calming, mood-enhancing effect in the brain. Because of this, amino acids as "mood foods" are now becoming part of a new medical frontier.

The nutritional value of a protein depends, in part on the percentage of its essential amino acids composition. In the AFA algae, essential amino acids compose an amazing *48.2 percent of its dry weight*, compared to only 30 percent in the soy bean protein.[3]

Essential Amino Acids

ARGININE is especially required by children and stress-affected people. Found abundantly within AFA algae's storage granules, arginine biostimulates the thymus gland and thus enhances our entire immune system. It stimulates white blood cells to fight infection and inhibit tumor growth and size. For this reason alone, AFA's arginine should be researched as an antitumor agent. Because arginine is not patentable, however, many drug companies have not fully pursued this line of research.

Ingesting AFA also helps to build up an "arginine reserve" in the liver, which can then biostimulate the activity of enzymes that synthesize liver detoxifying proteins. This keeps the liver healthy. Arginine even biostimulates the pituitary gland to secrete human growth hormone, which causes some muscle-building and fat-burning effects. It is for this reason that eating algae enables wounds and burns to heal much faster. Ingesting arginine also increases sperm production because human sperm is very rich in this amino acid. Arginine is also available from nuts, meat, and cheese.

HISTIDINE, like arginine, is a conditionally essential amino acid that is available from animal and plant protein. Not much is known about it other than its role in metabolic processes involving red and white blood cell production. Since histidine bonds to and chelates metal ions such as copper and zinc, it increases their intestinal absorption. Like arginine, histidine boosts the activity of white blood cells. Some people have reported that histidine is of benefit in the treatment of rheumatoid arthritis.

ISOLEUCINE is an essential branched-chain amino acid (BCAA) that is available in most animal and plant proteins. Concentrated mostly in our muscle tissue, it is biochemically needed in energy production. Isoleucine appears to be a muscle builder, especially for those who have been ill or need to restore lost muscle mass. It seems also to reduce muscle tremors, possibly by stabilizing blood sugar. There is evidence that isoleucine is useful in treating liver damage and the neurological side-effects from alcoholism.

LEUCINE is also an essential branched-chain amino acid (BCAA) that is available in animal protein and AFA algae. Similar to isoleucine, it is concentrated in muscle tissue and used to produce energy. It probably stabilizes blood sugar and is particularly helpful in reducing some of the uncomfortable symptoms of hypoglycemia.

LYSINE is an essential amino acid that is available in wheat germ, fish, dairy products, and AFA algae. Concentrated in muscle tissue even more than the BCAAs, lysine with vitamin B_6 synergistically helps bone growth by improving digestion, thereby aiding the intestinal absorption of calcium. Along with vitamin E and iron, lysine helps form collagen, an important skin protein. The fact that lysine deficiency suppresses the immune system helps to explain why it is possibly useful in the treatment and prevention of genital herpes simplex infections.

Lysine also aids in the prevention of osteoporosis because as a chelator it helps bones to absorb calcium and thus reduces its loss through the urinary tract. At the same time, it even chelates and reduces lead toxicity. An interesting article published in the *Saturday Evening Post*[4] describes how a man was able to overcome and virtually eliminate his chronic fatigue syndrome by ingesting lysine.

METHIONINE is the scarcest of the amino acids. The best source of methionine are nuts, sunflower seeds, rice, corn, liver, eggs, fish, and AFA algae. Its sulfur content makes it important for skin and nail growth, fat reduction in the body and liver. It also helps prevent fatigue and depression. When bioconverted to cystine, methionine also serves as an antioxidant. Methionine increases the bioproduction of specific enzymes, which, in turn, favorably increases the amount of memory-enhancing neurotransmitters, such as dopamine, and mood-elevating, morphine-like polypeptide molecules called endorphins. These molecules can be directly absorbed from AFA into the human bloodstream, thus positively and quickly affecting physiological behavior. Methionine also has lipotropic effects and acts as a detoxifier of the liver.[5] Unfortunately, some studies suggest that excessive alcohol destroys or otherwise greatly reduces the amount of methionine in the system.

Methionine was probably one of the first amino acids available in Earth's ancient primordial seas billions of years ago. This amino acid was (and is still) used by primitive bacteria and blue-green algae to biosynthesize glutathione, possibly Earth's first antioxidant (protection) tripeptide molecule. Methionine in this form has been shown to help humans detoxify lead and copper contamination in the blood. Some studies have shown radiation protection as well. Also note that some friendly bacteria (if we have a healthy intestinal flora) can synthesize – with the

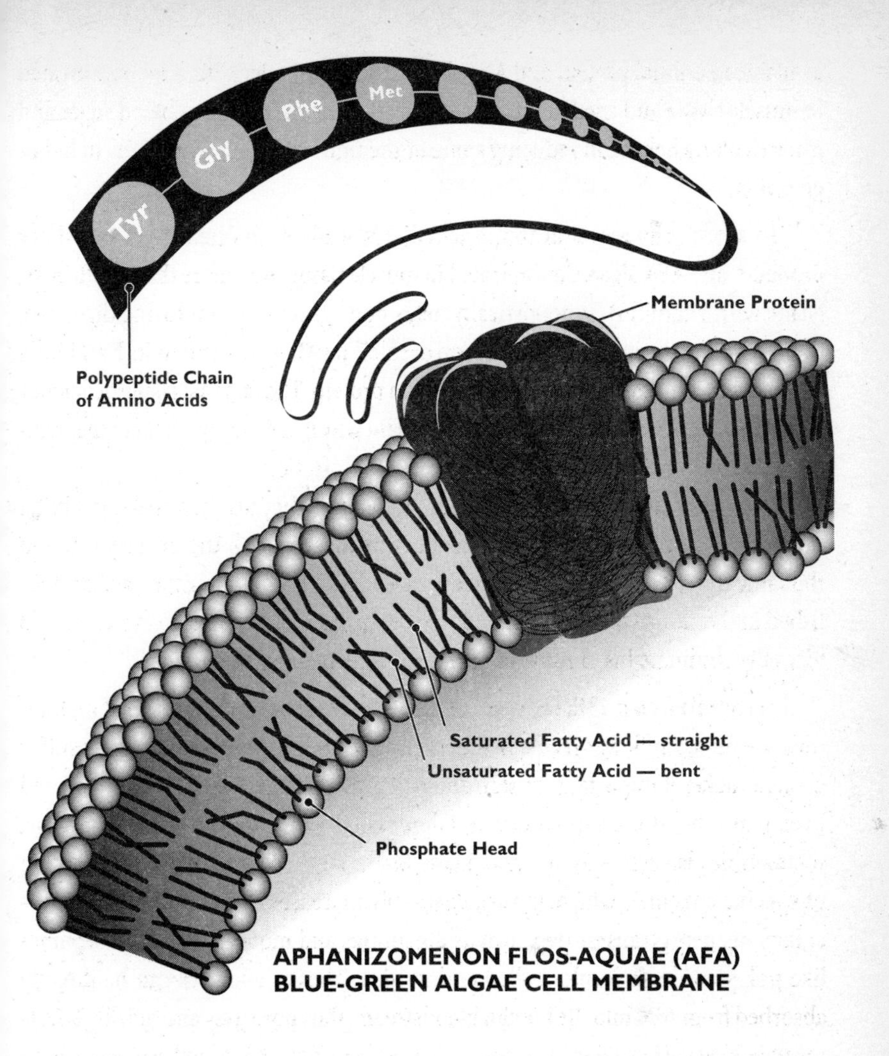

**APHANIZOMENON FLOS-AQUAE (AFA)
BLUE-GREEN ALGAE CELL MEMBRANE**

help of magnesium and vitamin B_{12} – some of our needed methionine from the more accessible aspartic acid.

PHENYLALANINE is an essential amino acid consisting of a phenyl ring of six carbons bonded to alanine, a smaller nonessential amino acid. Found in milk, meat, wheat germ, and AFA algae, this interesting molecule has a direct effect on brain chemistry because of its relatively rare ability to very quickly cross the blood-brain barrier. Phenylalanine can reduce some symptoms of depression, probably because it is able to prevent certain enzymes from breaking down mood-elevating endorphin molecules.

In the liver, phenylalanine is used to biosynthesize and increase the concentration of tyrosine, which is then further enzymatically converted to such important neurotransmitters as adrenaline and dopamine. Because of this, and with the help of B-vitamins, alertness, concentration, learning abilities, and resistance to addictive cravings for alcohol and sugar are noticeably enhanced.

Phenylalanine reduces back pain and arthritis (and even dental pain) by slowing down the breakdown of enkephalin and endorphin peptides. These peptides are morphine-like molecules that elevate mood and provide pain relief. Phenylalanine is found within a wide variety of brain and intestinal messenger peptides, which very quickly cross the blood-brain barrier.

Unfortunately, too many people drink diet sodas that contain the artificial sweetener aspartame, which is in the form of a curious dipeptide – phenylalanine with aspartic acid. As a result, too much phenylalanine is ingested, causing excessive absorption by the brain and liver. According to a study published in the *Journal of the American Medical Association*,[6] imbalances in the brain brought on by excessive and continuous aspartame consumption will contribute to memory loss. Although unhealthy for everyone, aspartame is especially dangerous for pregnant women.[7]

THREONINE, often called the "immune booster," is an essential amino acid found in dairy, meat, wheat germ, nuts, seeds, and AFA algae. It is especially needed for skin proteins like collagen and elastin and plays a minor role in reducing fat buildup in the liver. There is some evidence that threonine can enhance the immune system and promote healing by biostimulating the thymus gland to enlarge and produce more T-cell lymphocytes that are more active because of more flexible cell membranes. In experiments with mice, moderate deficiency of threonine produced a drastic depression in the immune function response to tumor growth. Threonine also boosts the immune system by stimulating the thymus gland.

TRYPTOPHAN is an essential amino acid that may be considered an essential vitamin because it is used in the biosynthesis of niacin (vitamin B_3). Tryptophan is a precursor for the very important neurotransmitter serotonin, a brain hormone and general messenger molecule, which is needed by the brain to bring on sleep and affect mood patterns. AFA's tryptophan changes in the body to form serotonin through the assistance of enzymes. Serotonin has other effects, such as blood vessel constriction, intestinal peristalsis promotion, and neurotransmitter capabilities for nerve cells. Serotonin also increases the permeability of sugar across the cell membrane. Tryptophan is found in small amounts in AFA algae, turkey, and milk, as well as some nuts and seeds.

VALINE is an essential branched-chain amino acid (BCAA) that is available in most foods as it is in AFA. Valine deficiency has produced "aimless circling" in laboratory animals. This could be caused if too much isoleucine and leucine go to the brain, blocking transport of valine.

Nonessential Amino Acids

ASPARAGINE is biosynthesized from aspartic acid. It is being studied in therapy for treatment of some types of depressed patients.

ALANINE is found in meat, cheese, wheat germ, and AFA algae. Concentrated in muscle tissue, alanine can be easily changed into glucose for muscle energy. It also helps to build muscle tissue. People with hypoglycemia may have an alanine deficiency. The alanine in AFA may be useful in treating hypoglycemia by stimulating an increase in blood sugar. (Alanine levels rise and fall with blood sugar in both hypoglycemics and diabetics.) Alanine is often found inside the polypeptides that make up cell walls. With the help of zinc and vitamin B_6, alanine biostimulates the immune system by increasing the size of the thymus gland.

GLUTAMINE is derived from glutamic acid and plays a major role in DNA synthesis. Glutamine is a neurotransmitter (like glutamic acid) and a "brain fuel." Its concentration in the blood is about 3.5 times higher than any other amino acid, and its concentration in the cerebrospinal fluid is even considerably higher. Manganese is needed for the proper metabolism of glutamine because of its presence in the enzyme glutamine synthetase. AFA is an excellent source of this mineral.

CYSTINE is an important sulfur-containing amino acid that is synthesized from smaller cysteine molecules in the liver and then used in a variety of interesting and powerful metabolic pathways. It actually takes two smaller cysteine molecules, each held to each other by a bridge of two sulfur atoms, to construct the larger cystine molecule. In addition, the shape and function of many large protein molecules depends, in part, upon sulfur to hold them together into elaborate three-dimensional patterns. For example, the shape of the protein gluteus, which gives wheat its elasticity, or the protein keratin, which makes tortoise shells hard and human hair curly, is attributed to these sulfur bonds between cysteine molecules.

Cystine is found in liver, onions, and wheat germ, and to a small extent in AFA.

Sometimes considered an "anti-aging" nutrient because of its antioxidant properties, cystine helps to detoxify carcinogens by directly using its sulfur atoms to absorb and stop the damaging effects of free radicals. To some extent, the resulting cystine molecules also help to increase hair growth and reduce balding because the keratin in hair must contain about 12 percent cystine to be healthy.

GLYCINE is important in brain chemistry as well as DNA, hemoglobin, and collagen synthesis. Found in protein foods and AFA, it is structurally the simplest amino acid. Glycine tastes like glucose (hence its name) and has a calming effect. Receptor sites for glycine are found throughout the central nervous system. Deficiency may lead to dementia and Alzheimer's disease.

Glycine has a variety of nutritional uses. It probably helps to lower cholesterol levels and triglyceride lipids in the blood. It stimulates the synthesis of glutathione, the most important endogenous antioxidant and detoxifier in all living organisms. And it even detoxifies benzoic acid, a common food additive.

Glycine also stimulates the secretion of glycogen, and therefore raises blood sugar and helps to alleviate fatigue. It is used orally and in ointments to heal wounds because skin collagen is a protein that requires high amounts of glycine. Glycine has been detected in star-forming regions outside of our own solar system, leading astronomers and biologists to suggest that extraterrestrial life exists.

PROLINE is readily found in meat, dairy, wheat germ, and AFA algae. Since it is a main component of collagen, proline is helpful in skin, tendon, and cartilage formation. About half of the body's proline is contained within collagen, leaving the other half distributed in almost every other protein in the body. As people age, the presence of urinary proline is a sign of collagen breakdown. Proline seems to help in wound healing, as does glycine, because proline increases collagen synthesis. Also, substance P, a neurotransmitter, is one of several small peptides that contain proline. As a result, proline peptides have neurological functions, some of which may possibly contribute to enhanced learning abilities.

SERINE is mostly found in meat, dairy, and AFA algae, and biosynthesized from glycine or threonine in the presence of B-vitamins. It is used primarily to construct brain proteins and nerve coverings. Serine is important in DNA and RNA synthesis as well as in the formation and stabilization of cell membranes. Serine is needed in the formation of the phospholipids of cell membranes and contributes to "membrane fluidity." It is good for the skin and has even been used as a natural moisturizer in skin creams. Unfortunately, too much serine from poor-quality foods (e.g., lunch meat and sausage) can suppress the immune system. The serine-containing enzyme, serine hydroxyl-methyl transferase, is in an unusually low concentration in psychotic patients. Ingesting good-quality serine may possibly reverse this, and more research needs to be done in this area!

The "net protein utilization" of AFA – an index of how well amino acids can be assimilated by humans – has been measured at a value of 75%, while that of SPIRULINA and CHLORELLA, two other edible algae, are only 37% and 20%, respectively. Red meat is surprisingly low, at only 18%. During times of stress and disease these amino acids need to be utilized as efficiently as possible.

TYROSINE is concentrated mostly in muscle tissue; it occurs a little in the brain. Tyrosine is biosynthesized from and biochemically similar to phenylalanine. It is known as the "anti-depressant" amino acid because of its importance to brain nutrition. Mountain climbers and U.S. soldiers have consistently reported less soreness, and fewer headaches and colds during tyrosine studies. Tyrosine has been used as a drug detoxifier and antioxidant and has been very successful in treating cocaine and substance-abuse withdrawals.

Tyrosine has biostimulating properties and is a precursor of the neurotransmitters' adrenaline (epinephrine), norepinephrine, and dopamine, as well as the thyroid hormones. A subtle decrease in the amount of thyroid hormones is immediately connected to overt changes in mood, quite often leading to states of depression.[8] The only substantial source of tyrosine in food is from quails, peanuts, and AFA algae.

Dopamine is especially important in that it promotes mental alertness, learning, and memory. It reduces such premenstrual stress (PMS) symptoms as anxiety, irritation, mood swings, and fatigue. Dopamine acts as a pain reliever and sometimes reduces the appetite.

ASPARTIC ACID is closely related to asparagine. You may recall that cyanophycin storage granules in AFA are made of aspartic acid and arginine. Small amounts of aspartic acid are known to biostimulate and increase the weight of the thymus gland in laboratory animals, enhancing their ability to resist infection.[10]

Such biostimulation will also lead to an increased production of white blood cells, which gently boosts the immune system to resist cancer. Aspartic acid is found in a variety of protein food and has been useful in treating chronic fatigue syndrome partly because of its involvement in DNA/RNA synthesis and partly because it acts as a neurotransmitter. This amino acid has a higher concentration in the human brain than any other amino acid. Research has shown that salts of aspartic acid, which are present in AFA as potassium aspartate and magnesium aspartate, are useful for protection against radiation damage. Because of its ability to chelate radioactive ions, blue-green algae is being used to help treat radiation victims in Chernobyl.

GLUTAMIC ACID is a "brain fuel" that is readily obtainable from animal, vegetable, or blue-green algae. It is concentrated in the human brain's memory center as a "stimulant" or excitatory neurotransmitter. With the help of vitamin B_6 and manganese, glutamic acid is enzymatically biosynthesized into GABA (gamma-amino butyric acid), another important inhibitory neurotransmitter present throughout the central nervous system. GABA is believed to reduce alcohol and sugar cravings.

There are a number of interesting claims about this amino acid. Megadoses of glutamic acid have been shown to increase IQ and mental functioning. More thoughtful studies are required.

The smaller concentration of glutamic acid in AFA also acts as a neurotransmitter to influence memory and mental acuity by stimulating brain cell receptors.

Powerful Polypeptide Antioxidants

Amino acids bond to one another to form polymer chains called dipeptides, tripeptides, or polypeptides. AFA contains a tripeptide called glutathione (GSH), which was most likely Earth's first protective antioxidant molecule. It was probably used by primitive forms of bacteria, and it is still used by all blue-green algae. When we ingest AFA, the GSH tripeptide not only serves as a valuable antioxidant, but it also helps detoxify our blood. It does this by wrapping around and "chelating" environmental toxins such as lead, mercury, and cadmium ions so that they may be more easily excreted.

AFA algae is a rich source of polypeptides, some of which are saved for later use in specialized storage granules. Even part of the soft cell wall of AFA is made of short and digestible polypeptides. Once ingested, they are biomodified as needed by removing one or another functional group.

Enzymes – Movers and Shakers of Life

Enzymes are specialized protein molecules that act as catalysts to speed up the biochemical reactions within any living cell by setting a stage upon which reactions can take place more quickly and efficiently. This occurs only when the enzyme acts as the biochemical key that unlocks the reacting molecule by distorting its shape so that its bonds break just as they should.

The cellular activity of all life depends upon a symphony of such specific enzyme reactions. Being among the most independent of all unicellular organisms, AFA is able to efficiently use its specialized enzymes for a variety of purposes. One among these is to change atmospheric nitrogen into needed amino acids for growth and repair.

All life makes use of enzymes such as the *transaminase* enzymes to *transform* one kind of amino acid into another, or digestive enzymes which break the bonds that hold together molecules of proteins, fats, and carbohydrates. There are six different types of enzymes, performing a variety of specific catalytic functions:

- *Transferases, which move functional groups*
- *Hydrolyases, which break bonds (using water)*
- *Isomerases, which rearrange atoms in a molecule*
- *Redox enzymes, which remove or replace electrons*
- *Ligases, which connect small molecules to each other*
- *Lyases, which break bonds (without water)*

Because of these many functions, we may think of enzymes as literally the "movers and shakers" of biochemistry.

Most enzymes require a smaller bonded coenzyme to function properly. The coenzyme, or that part of the coenzyme that humans cannot biosynthesize, is called a vitamin. Enzymes typically have their vitamins bonded or "chelated" by one or more amino acids. Many of these enzymes require vitamin B_6 as a coenzyme. Chelated vitamins are much more easily assimilated by humans than unchelated inorganic sources.

Most enzymes require the presence of one or more metal ions as "cofactors" to help the enzyme function and keep its shape. Zinc, magnesium, and iron are such necessary cofactors for hundreds of enzyme systems. Superoxide dismutase (SOD) is an enzyme that AFA uses to protect its DNA from dangerous "superoxide" free radicals. SOD requires copper, zinc, and manganese as mineral cofactors. By ingesting AFA's enzymes, we are also being nourished by a wide range of its chelated vitamins and minerals, as well as the enzymes and amino acids that compose it.

DNA and RNA – Nucleic Acid Brain Food

Toward the center of AFA's cellular cytoplasm, and unbound by any nuclear membrane, is one incredibly long DNA molecule, which carries all of AFA algae's genetic information. Unlike the more advanced eukaryotic cells, AFA's single and circular DNA strand is not entangled with any complex proteins. This is perhaps one reason why the nucleic acids of the DNA can so easily provide the mental clarity enhancement so often reported by algae eaters.

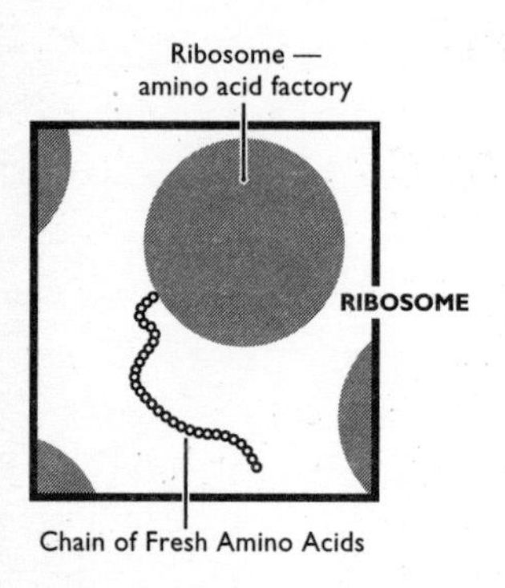

Also scattered within the cell are mysterious protein-making enzyme complexes called ribosomes. Each is thought to consist of several smaller proteins bound to a unique nucleic acid called ribosomal RNA. The exact structure of these ribosomes remains a mystery, but one thing is certain: The DNA and RNA dictate which amino acids are actually connected together to form proteins. These same amino acids and neuropeptides of AFA, along with its DNA and RNA, form a powerful team that most likely enhances brain function.

Ribosomes, scattered by the thousands throughout the cytoplasmic interior of AFA cells, are the cellular factories within which life-sustaining proteins, enzymes, and polypeptides are manufactured so the cell can properly function. One enzyme (nitrogenase) is used to process nitrogen into amino acids, while others are used continually to repair the growing cell wall. Polypeptides are needed to protect the delicate complexities of DNA and RNA. Any damage to them will cause the manufacture of erroneous enzymes that cannot perform their intended function. Other proteins used as "channels" are found embedded within the cell membrane as gate keepers to allow nutrients in and toxins out. In the next chapter, we take an in-depth look at how the efficiency of the cell membrane of AFA algae dramatically depends on the fluidity created by special and essential fatty acids.

CHAPTER FOUR

The Celebrated Cell Membrane of AFA Algae

APHANIZOMENON FLOS-AQUAE CELLS ARE RELATIVELY small barrel-shaped cells that are actually about 25 times smaller than the cells which comprise the food we usually eat. Because of their size, prokaryotic AFA cells have about 25 times more surface area than the larger eukaryotic cells. To the algae eater, this simply means there are more micronutrients available in their membrane. One important substance found in the cell membrane are the fatty acids: they give it life-sustaining flexibility.

Fatty Acids – Three Families

Remember lipids – sometimes called fats – naturally divide into a spherical, water-soluble "head" and two long, water-insoluble "tails." Cell membrane lipids are referred to as phospholipids due to the presence of phosphorous in the head.

Each of the two hydrocarbon tails, which attach to the head of every phospholipid, are actually fatty acids. Their acid functional groups are hidden within the water-soluble polar head. Typically, all fatty acids are in this form.

The most common and abundant fatty acids in all cell membranes are those with an even number of carbons, usually 16 or 18. The shape of the fatty acid tail is critical to the health of the cell and that depends on whether the tail is saturated, monounsaturated, or polyunsaturated.

- SATURATED *fatty acid tails are somewhat straight because there are no double bonds that cause the molecule to curve. Such fatty acids are not good for the cell membrane because they impart an unhealthy rigidity to it. Rats fed diets of the saturated fatty acids found in coconut oil and milk fat are known to develop irregular heartbeats.*[1] *AFA algae's membrane has one of the lowest amounts of saturated fatty acids and contains absolutely no cholesterol. Dietary saturated fatty acids are often peroxidized to the point of being slightly rancid. When such fatty acids are incorporated into the cell membranes of otherwise healthy cells, the structure and function of the cell is considerably impaired.*
- MONOUNSATURATED *means that the hydrocarbon tail is slightly bent due to the presence of one double bond. As a result, monounsaturated fatty acids help the cell membrane to be a little more flexible than the saturated types. According to the Journal of the American Medical Association,*[2] *non-insulin diabetics need monounsaturated fats (available in olive oil, almond oil, and the oils of AFA) to help reduce their high levels of low-density lipids (LDLs, also known as "bad cholesterol") and prevent the blood vessel damage and deterioration that results from poor glucose-insulin balance.*
- POLYUNSATURATED *means that the hydrocarbon tail is bent in several places because of more than one double bond. These fatty acids help the cell membrane to be much more flexible than the saturated or monounsaturated types. Soybean, sunflower, and sesame oils, as well as the oils of AFA, are good sources of our very much needed polyunsaturated fatty acids (PUFAs), which account for 25 percent of human brain weight.*

According to a study published in *The Lancet*,[3] human breast milk contains a high concentration of many of the polyunsaturated fatty acids needed for the brain development of infants. When infant formula is supplemented with PUFAs there is a dramatic improvement in the brain development of those infants who are fed formula and are not breast fed. Additional research needs to be done to see the effect of AFA algae on formula-fed infants.

Although unsaturated fatty acids enhance the flexibility of all cell membranes, their unsaturation also makes them particularly vulnerable to the oxidizing effects of incoming free radicals that all too often lead to the production of "rancid" fatty acids. Antioxidants in AFA, such as betacarotene, superoxide dismutase (SOD), and GSH enzymes, inhibit this complex process so that the cell membrane does not become inflexible and unhealthy. In addition, evidence shows that having a wide variety of antioxidant carotene compounds (as those available in AFA) dramatically helps to maintain the flexibility of our respiratory tract cells.[4]

Unfortunately, the peroxidized fatty acids from dietary fats and oils interfere with our normal cell function by altering the structure of the cell membrane.[5] Our diet should contain vegetable oils such as linoleic and linolenic acids. AFA algae has high amounts of both, and the antioxidants to defend them from free radical attack.

The Cell Membrane – Central to Good Health

In the primitive oceanic environment of ancient Earth billions of years ago, fatty acids first coalesced into what we now call the cell membrane. The laws of biophysics seemed to demand that the cell membrane exist first, so that a central replicating DNA molecule could soon follow. This began that important and necessary separation between the dangerous outside world and a new life form protected and evolving inside the membrane of blue-green algae.

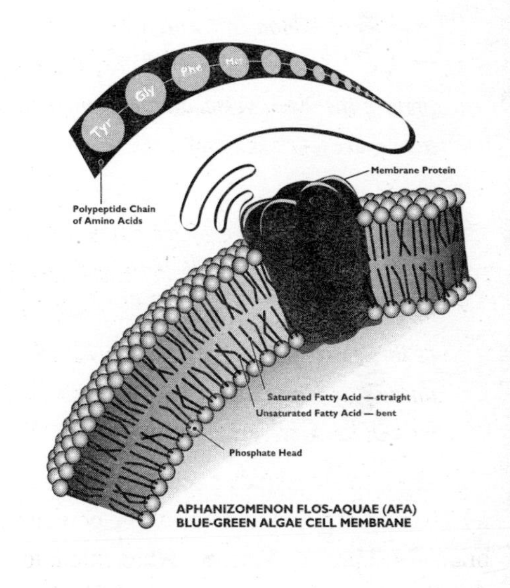

APHANIZOMENON FLOS-AQUAE (AFA)
BLUE-GREEN ALGAE CELL MEMBRANE

The most interesting and nutritionally valuable portion of the AFA cell may be its thin outer covering called the cell membrane. This critical barrier surrounds the cell and consists of two layers of phospholipids distributed such that the tails of both layers point towards each other. Their water-soluble polar heads form a kind of cobblestone matrix on the outside and inside layers.

In addition, the cell membrane contains protein molecules embedded between the phospholipids. Some of the proteins actually traverse both sides of the membranes, whereas others are located more on the outside or inside regions. Also connected to the inner cell membrane are various enzymes, which continually repair and expand the framework of the surrounding cell wall.

Imagine now, for more complex human cells, there are "whiskers" or chains of interconnected sugar molecules attached to and protruding off of some of the protein molecules on the cell membrane surface. These antenna-like whiskers, sometimes called *receptor sites*, receive outside information from hormones and neurotransmitters secreted by other cells.

Usually these receptor sites are made of chains of sugar molecules. For example, hormone messenger molecules like adrenalin attach to these receptor sites on the membrane's outer surface and cause other internal messenger molecules to activate specialized proteins that pump important ions into or out of each cell. In a similar manner, neurotransmitter molecules also attach to protein receptor sites and trigger the transfer of crucial molecular information among our brain cells.

Phospholipids and Membrane Fluidity

The fluidity of the cell membrane is important to the health of any cell. Whether across the phospholipid bilayer, or through the tunnel-like holes of the imbedded proteins, ions and molecules of nourishment and waste must be able to ceaselessly and gracefully flow.

In general, saturated fatty acids tend to pack together tightly and inflexibly. Saturated fats are also dangerous because they increase damage to the artery walls.[6] On the other hand, unsaturated fatty acids always provide a kind of "springiness" to the cell membrane. This promotes flexibility, fluidity, and ultimately greater health to the cell. Polyunsaturated fatty acids also have these beneficial effects.[7]

AFA algae has an unusually high amount of such unsaturated fatty acids. By eating AFA, a generous amount of its healthy and unsaturated fatty acids can be incorporated into the cells of our body. This helps skin cells to be more resilient and able to detoxify. Similarly, white blood cells of our immune system are strengthened by such membrane fluidity, which allows them to more efficiently maneuver in our bodily fluids and better attack viral and bacterial invaders.

Usually each phospholipid unit within a cell membrane will have, on the average, one saturated and one unsaturated fatty acid. The blue-green algae in general have an unusually high percent of polyunsaturated fatty acids. AFA in particular exemplifies this trend with an exceptionally high concentration of such polyunsaturated fatty acids.

AFA has an advantage in being hundreds of times lighter than the cells of plant and animal food. Its high surface area allows for cell membrane material to accumulate nutrients and eliminate waste and even use the chlorophyll embedded within the cell membranes to receive more energy from the sun. This makes AFA among the most self-sufficient organisms on Earth with an unusually high concentration of valuable nutrients. AFA produces more cell membrane material without getting larger by creating a vast system of membrane invaginations similar to the brain's infoldings. Having such a rich cell membrane is why AFA has such a high amount of chlorophyll, protein, and essential fatty acids.

Essential Polyunsaturated Fatty Acids – Absolutely Essential

Essential polyunsaturated fatty acids are polyunsaturated fatty acids (PUFAs) that cannot be biosynthesized by humans – that is, we must eat them. Collectively they are called essential fatty acids (EFAs), or vitamin F. They come primarily from vegetable oils, seeds from borage and evening primrose, and the cell membranes of blue-green algae such as AFA. They are potent biomolecules and sometimes wield a powerful biological response. EFA deficiency has been known to substantially lower the number of thymus suppressor cells and thus impair the immune system.[8]

Essential fatty acids, like those present in AFA algae, are also required for the normal growth and repair of the skin, blood vessels, and nerve tissues. They are vital for efficient respiration and the fluidity and flexibility of all tissues and cell membranes, particularly in the retina and the brain. EFAs have lubricating qualities. They are known to reduce blood cholesterol and thus help to prevent cardiovascular disease. *Time* magazine has reported[9] that many heart patients suffer from a deficiency of EFAs and should be ingesting them for heart disease prevention. EFAs are especially useful because of the efficiency with which they increase the solubility of cholesterol deposits and wash these deposits away from our artery walls.

A deficiency of essential fatty acids can cause anorexia, hyperactivity, arthritis, acne, dry skin, hair loss, diarrhea, and slow wound healing. AFA, cod liver oil, and coldwater algae-eating fish such as salmon, sardines, and mackerel supply polyunsaturated fatty acids that can compete with and diminish the deleterious effects of arachidonic acid (AA) in animal fat. This reduces inflammation[10] and often the pain of rheumatoid arthritis.

The essential fatty acid eicosapentenoic acid found in fish oils and blue-green algae also seems to inhibit the production of inflammatory leukotrienes, and thus psoriasis is diminished.[11] As the consumption of fish oils or essential fatty acids found in AFA increases, the tendency for blood platelets to aggregate and increase blood pressure goes down. A positive result of this is a less frequent occurrence of painful migraine headaches.[12] Another reason for ingesting AFA is that it may help reduce PMS symptoms that are associated with a shortage of essential polyunsaturated fatty acids and an excess of the female hormone prolactin.[13] The *New England Journal of*

. . . .

The Omega-6 Group

LINOLEIC ACID (LA) is an essential fatty acid and a major constituent of flaxsee oil and many vegetable oils, as well as the AFA cell membrane. Lack of linoleic aci in laboratory animals leads to scaly dermatitis.[17] When linolenic acid is increase heart attack rates are decreased.

GAMMA-LINOLENIC ACID (GLA) is a semi-essential fatty acid that is criti cal to the health of the cell membrane. It is enzymatically biosynthesized from linolei acid by the enzyme delta-6-desaturase (along with zinc and vitamin B6). The desat urase portion of the enzyme name tells us that that a more unsaturated fatty acid in the making. However, this enzyme is inhibited by some of the fatty acids found i milk and animal products, as well as by alcohol consumption.[18]

GAMMA-LINOLENIC ACID is present to varying extents in evening primrose o (7–10%), borage oil (18–26%), spirulina blue-green algae (0.7%), and AFA alga (28%). The GLA in both evening primrose oil and AFA effectively lowers high choles terol levels, with GLA being about 170 times more potent than the LA portion of th Omega-6 group.[19] GLA is also needed to inhibit platelet aggregation.

Medicine has reported[14] that our sensitivity toward insulin is probably connected to the number and concentration of polyunsaturated fatty acids in the cell membranes of skeleton-muscle cells. Cell membrane PUFAs determine the sensitivity of a cell membrane to insulin. The higher the polyunsaturated fatty acid concentration within the membrane, the higher the insulin sensitivity.

Essential fatty acids consist of two groups: (1) linoleic acid and gamma-linolenic acid in the so-called Omega-6 group; and (2) alpha linolenic and eicosapentaenoic acid in the Omega-3 group. The "6" of Omega-6 means that the first double bond in the molecule is located six carbons away from the beginning of the nonpolar tail. Likewise, the "3" of Omega-3 means that first double bond is three carbons away. Many dietitians feel the human body needs twice as much of the Omega-3 fatty acids as the Omega-6 because the former are known to lower LDLs, the "bad" cholesterol that leads to heart disease. The proportions of these fatty acids in AFA algae are very close to this ratio.

. . . .

The Omega-3 Group

ALPHA-LINOLENIC ACID (ALA) is an essential fatty acid and major component of flaxseed, many vegetable oils, and especially AFA. Alpha-linolenic acid is a precursor to another group of prostaglandin molecules called PGE3. According to an extensive study reported in the medical journal The Lancet,[20] diets rich in ALA are very effective in preventing further attacks in heart attack patients.

EICOSAPENTAENOIC ACID (EPA) is a semi-essential fatty acid present in cold-water fish such as cod, mackerel, and salmon, and, of course, AFA algae. Greenland Eskimos have greatly reduced cardiovascular disease because they eat such cold-water EPA-containing fish. These Eskimos also experience a lower incidence of rheumatoid arthritis, hypertension, psoriasis, high blood pressure, and atherosclerosis. Of course, these cold-water fish are known to dine on large quantities of blue-green algae. Eicosapentaenoic acid treatment may even decrease lung damage in patients with cystic fibrosis, according to The Lancet.[21] Patients treated with 2.7 grams of EPA per day for six weeks experienced a dramatic decrease in sputum production.

Mother's milk is that precious fluid created to uniquely prepare us for our long and special human journey. One of its most powerful micronutrient components—gamma linolenic acid (GLA)—is so precious that it is found within only a few rare and treasured natural sources.

One such precious source that also has a similarly special evolutionary journey is AFA. Roughly 10 percent of the weight of dry AFA is gamma-linolenic acid. It takes only a small amount of GLA to bring on profound and important physiological changes. GLA is used in the body to release neurochemicals that can improve mental attitude and reduce depression, help in weight loss, and simultaneously improve skin tone. Even enhanced mental clarity and alertness, along with a general "up feeling," may be realized. In the form of evening primrose oil, back in merry old England, it was called the "King's cure-all."

Not all cholesterol is bad, however. Cholesterol is important in that its somewhat flat and polycyclic molecules are designed to stack between the fatty acid molecules, helping to hold them together and give integrity and strength to the cell membranes. The drawback is that too much cholesterol holds the membrane together too rigidly so that the membrane loses its fluidity and thereby its health.

Also, when excessive cholesterol builds up within a human cell membrane, it sometimes breaks down into smaller pieces (metabolites), which tend to accumulate within prostate cells, irritating and enlarging them.[15] This process is accelerated by a dietary deficiency in essential fatty acids.[16] There is no cholesterol in prokaryotes such as AFA algae.

The Omega-3 fatty acids present in AFA help to make the blood less sticky and reduce clot formation.[22] According to the University of California at Berkeley Wellness Letter,[23] fish oil supplements may not be the safest way to get our essential fatty acids. Ingestion of AFA on a regular basis is strongly recommended as an alternative.

The *New England Journal of Medicine* has reported[24] that polyunsaturated fatty acids found in fresh fish may help protect cigarette smokers against chronic obstructive lung disease. Results of another study published a few months later in the same medical journal[25] suggest that natural fish oil PUFAs can even "slow down the deterioration of kidney function." The essential fatty acid content of AFA ought to be comparatively effective. This is another opportunity for further research.

Several studies have shown that an increase in dietary fish oils can lessen the painful effects of gout.[26] The EPA – eicosapentaenoic acid – in fish oil limits the production of leukotriene molecules and thus tends to bring down inflammation.[27]

Science News has reported[28] that a study by Hilbben and Salem links a deficiency in decosahexaenoic acid (DHA) to depression. Seafood and AFA are primary sources of this polyunsaturated fatty acid because it can be biosynthesized readily from EPA of the Omega-3 group. The rather remarkable relationship between mood and AFA algae ingestion is treated in Chapter 9. *The Lancet*[29] also reports that DHA is important in the development of both the nervous system and visual acuity in newborns.

Lecithin – Abundant in the AFA Cell Membrane

Lecithin, the most common phospholipid, is present in and important to the structure and flexibility of all cellular membranes – plant, animal, or algae. Its non-polar tail consists of two mostly polyunsaturated fatty acids. Because its polar head consists partly of phosphate and choline, it is also named phosphatidylcholine.

Lecithin is known to increase cholesterol solubility and helps to remove it from clogged arteries. This reduces the dangers of atherosclerosis due to lipids accumulating in the arteries.[30] Unfortunately, the lecithin found in egg, as compared with that in AFA, consists mostly of saturated fatty acids and does not show this effect.

The choline in lecithin is a B vitamin that is effective in maintaining cell membrane fluidity. Because of this, lecithin is believed to improve short-term memory and reduce the effects of early Alzheimer's disease.

Prostaglandins – Messenger Molecules

Essential fatty acids sometimes function as biochemical precursors in the synthesis of specialized biomolecules called prostaglandins. These are somewhat mysterious hormone-like chemical messenger substances that regulate and modulate, even in minute quantities, a variety of important cell functions. Prostaglandins are secreted by one cell to regulate the activity of another. A relatively large number of such hormones have already been discovered. Only a few of them are understood very well in terms of their biochemical response mechanism. Prostaglandins were originally given their name because of their association with research being done on the human prostate gland. They are actually found throughout the body, although they are very short-lived and are often immediately destroyed by the oxygen from the lungs.

An imbalance in the production of these hormone-like molecules, along with iron and vitamin A deficiencies, may actually be the cause of excessively heavy menstrual periods (menorrhagia). According to an article in *Science*,[31] prostaglandins are being carefully synthesized in the laboratory to better study their anticancer and antiviral therapeutic possibilities.

Aspirin is effective as a pain reliever because it prevents a specific enzyme (prostaglandin synthetase) from chemically changing a polyunsaturated fatty acid (arachidonic acid) into a prostaglandin molecule that regulates the perception of pain. As another dietary PUFA (linoleic acid) rises, blood pressure begins to go down due to an increase in the so-called E-series prostaglandins.[32] Since aspirin will retard such needed prostaglandin production, some physicians do not recommend its use.

As reported in *The Lancet*,[33] people with asthma should greatly diminish eating animal products. The arachidonic acid that ends up in our tissues is almost all derived from eating animal products.[34] This acid produces molecules (leukotrienes) that trigger allergic reactions and have an important role in water retention (edema) and puffiness. These molecules are probably 1000 times more problematic than histamine toward asthmatic bronchial constriction.[35]

Cis- and Trans-Fatty Acids

All unsaturated fatty acids have at least one double bond between at least one pair of their carbon atoms. The monounsaturated types have one double bond, while the polyunsaturated types have two or more. All of their double bonds may be described as being either cis or trans, depending on the geometric shape of their hydrocarbons. The cis-fatty acids are healthier to ingest because they are more curved in shape and contribute to the flexibility of the cell membrane. The trans-fatty acids, on the other hand, are less healthy because they are straighter in shape and able to be packed together with much less flexibility inside the cell membrane. Cis-unsaturated cooking oils (mono or polyunsaturated) are healthier than the trans or the saturated varieties of cooking oils for this reason.

For the very same reason, AFA algae is good for skin, heart, and brain tissue because it too is particularly high in cis-unsaturated fatty acids. Foods that are high in trans-fatty acids are milk, margarine, and hydrogenated vegetable oils. These should especially be avoided by those concerned with acne or premature facial skin wrinkling.[36] The trans-fatty acids in margarine and butter often increase cholesterol levels.[37] The polyunsaturated fatty acids in AFA seem to reduce some of these negative reactions. Laboratory mice are less likely to get breast cancer if they eat diets enriched with curved and flexible linoleic acid molecules.[38]

According to the research efforts of Zvi Cohen at The Laboratory for Microalgal Biotechnology in Israel and Helen Norman at the USDA, there is great medical importance in algae-derived Omega-3 PUFAs. If their consumption is kept high – relative to the Omega-6 PUFAs typically exaggerated in Western diets – human cell membrane flexibility will increase, while blood viscosity will decrease. As a result, there should be a lower incidence of arthritis, heart disease, atherosclerosis, and high blood pressure, as well as less breast, prostate, and colon cancer. AFA has a very high concentration of such PUFAs when compared to other plants, seeds, and nuts as well as other blue-green algae. The reason Upper Klamath Lake algae is more concentrated with PUFA is due to the fact that it grows in a relatively cold environment compared to that of tropical algae like *Spirulina.* AFA compensates for the cold by manufacturing more of the flexible Omega-3 PUFAs.[39]

THE FLEXIBILITY OF ANY CELL MEMBRANE is directly proportional to the number of polyunsaturated fatty acids that compose it. AFA's natural membrane fluidity imparts these qualities to each and every one of our cells. Brain and heart cells are especially appreciative because their optimum efficiency increases with their ability to be correspondingly supple and resilient. Even white blood cells benefit because the membrane flexibility allows them to quickly and efficiently attack invaders in otherwise inaccessible places within the body. It is no wonder that some biochemists have "knighted" these essential fatty acids by referring to them as "vitamin F." We like to think that the "F" stands not for "fatty" but for "flexible."

But the expression of a well-made man appears not only in his face,

It is in his limbs and joints also, it is curiously in the joints of his hips and wrists,

It is in his walk, the carriage of his neck, the flex of waist and knees . . .

WALT WHITMAN

"I Sing the Body Electric" FROM *Leaves of Grass*

CHAPTER FIVE

The Wide Range of Vitamins

PROPER FUNCTIONING OF MOST ENZYMES IN ANY ORGANISM is dependent upon a variety of smaller molecules called coenzymes. Any portion of the coenzyme that the human body is incapable of biosynthesizing for itself may be called a vitamin. Such vitamins are available by ingesting meats, vegetables, fruits, legumes, and AFA algae. It also helps to maintain a healthy gastrointestinal tract of friendly, vitamin-producing microflora.

The twelve available vitamins in AFA algae work together to vitalize and direct the internal enzyme systems so that virtually each and every biochemical transformation can take place properly. It is as if the enzyme molecule were a biological machine in a cellular assembly line that can work effectively only if skillfully operated by a vitamin worker molecule.

When a vitamin is organically attached or chelated to an enzyme, intricate molecular manipulations are made possible. Functional groups like alcohols and amino acids can now be removed from or added to whatever target molecule or substrate the enzyme-vitamin system is working on or transforming.

The B vitamins in AFA are easy to absorb and assimilate because they are organically chelated or bonded within various enzyme systems. This allows a small amount to have powerful effects like improving mental clarity. A number of other nutrient factors in AFA also do this. Perhaps the synergistic affect of all of them together may help to explain why so many people speak of dramatic mental upliftment after regular amounts of AFA. Without natural organic vitamins we would find ourselves physically and emotionally drained of energy.

Thiamine (Vitamin B_1) – Champion of the Nervous System

Thiamin (vitamin B_1) is the "nerve and energy" B-vitamin molecule obtained from brown rice, AFA, and other grains. By activating specialized enzymes, it helps to convert blood glucose into energy so that nerve, heart, and muscle tissue reactions can take place. This is why a deficiency of vitamin B_1 produces fatigue and even mental confusion. This vitamin deficiency was once called beriberi (meaning "I cannot move") by incapacitated Japanese sailors.

Thiamine is an especially useful coenzyme because when it is part of larger enzyme systems, it is able to alter the molecular structure of certain amino acids by removing their acid functional groups. This is a preliminary step in changing amino acids into hormones, which are then used by nerve cells for communication purposes. This is one reason why vitamin B_1 is so good for the health of the entire nervous system and so helpful toward alleviating many nervous system disorders. Studies show that B_1 helps to prevent lead, a ubiquitous pollutant, from depositing in nerve and brain tissue. As reported in the *Journal of the American Medical Association*,[1] children with high levels of lead in their blood are known to exhibit violent behavior.

Riboflavin (Vitamin B_2) – Enzyme Animator

Riboflavin is mainly an "antioxidant" vitamin that enlivens several enzyme systems, one of which helps to biosynthesize glutathione, an important tripeptide that protects against free radical damage.

When two riboflavin molecules (along with two phosphates) are bonded to a protein, a powerful enzyme system is created that is capable of removing tiny hydrogen atoms from other molecules so delicately held in its catalytic grip. This so-called "dehydrogenation" mechanism is essential for the synthesis of DNA, amino acids, and unsaturated fatty acids.

Since all of these mechanisms are used so often for cellular growth, riboflavin needs to be continually replenished. Deficiencies are very common, especially in the elderly and in alcoholics. Cracks in the corners of the mouth or on the lips are indicators of a deficiency in this vitamin. Fortunately, migraine headache severity drops for most people who regularly get their riboflavin.[2] Also, people who exercise a lot seem to benefit even more from riboflavin's antioxidant protection. Riboflavin is readily obtained from organic leafy green vegetables and AFA algae.

Pyridoxal (Vitamin B_6) – Immune System Booster

Vitamin B_6 is the "immune system booster" that is obtained primarily from whole grains, brewer's yeast, and AFA. Because this vitamin vitalizes over sixty enzyme systems, it is able to play a key role in the production of red blood cells and immune cells. The vitamin B_6 in AFA also biostimulates growth so that wounds seem to heal faster.

Vitamin B_6 is considered one of the most diversified coenzyme known. It specializes, when chemically part of the right enzyme system, in changing and transforming one amino acid into another. It does this by carefully moving, like a skilled craftsman, acid and amino groups from one molecule to another. Without vitamin B_6, proteins could not be constructed, and amino acid transformations such as that of serine into glycine could never take place. Since red blood cell production and DNA synthesis all require the biosynthetic skill of B_6, humans tend to feel alert and energized when it is available. Vitamin B_6 has even been known to help correct the tryptophan metabolism defects found in asthmatic children.[3]

Vitamin B_6 is an important coenzyme in the biosynthesis of compounds (monoamines) that are typically low in depressed patients.[4] AFA provides the vitamins and the amino acid building-blocks for such monoamine synthesis.

Niacin (Vitamin B_3) – Vanquisher of Tension

Niacin is the "stress reducer" vitamin molecule obtained from meat, fish, AFA algae, and leafy green vegetables. It is used in enzymes which convert food into energy, repair cells, and promote good health in general. Niacin was one of the first coenzymes to be recognized. Although AFA is relatively low in niacin, the amino acid tryptophan can be enzymatically converted to niacin with the help of B_6 to meet our dietary needs.

A lot of research points to niacin's considerable cholesterol-lowering effect, even in relatively small amounts. In the form of nicotinic acid, it seems to be able to reverse some atherosclerosis. According to one study, however, treating high blood cholesterol with the time-released niacin found in many inorganic pharmaceutical vitamin supplements may actually increase the risk of liver damage.[5] Natural food sources such as AFA algae provide the safest way to get vitamin supplementation.

Pantothenic Acid (Vitamin B_5) – Fatigue Fighter

Pantothenic acid reduces stress and fatigue by helping the adrenal glands function more efficiently. It is obtained from liver, eggs, AFA, and whole grains, and plays a variety of helpful roles in enzymatic reactions throughout the body. Vitamin B_5 is also an anti-aging vitamin that works as a coenzyme to enhance skin and nerve cells by repairing their damaged cell membranes. Vitamin B_5 is also an antioxidant vitamin that helps to protect cells from the oxidizing damage of free radicals. Because of such antioxidant properties, vitamin B_5 is sometimes helpful in alleviating some of the common pains of morning arthritis. It even strengthens the immune system by increasing the production of "natural killer" cells.

There are similarities as well as differences among the thousand or so species of blue-green algae. AFA and SPIRULINA both can biosynthesize the vitamin C they need, but AFA has about five times more vitamin C than SPIRULINA.

Folic Acid – Nutrient for the Intestines

Folic acid derives its name from the word *foliage* and may be obtained from a good diet that includes fresh kale, spinach, and AFA. In general, people do not get enough of this vitamin. In fact, folic acid deficiency may very well be *the most common vitamin deficiency in the world*, mainly because folic acid is so easily destroyed by cooking. Forgetfulness, insomnia, anemia, irritability, and even dementia may accompany a low concentration of this vitamin.

Folic acid also needs to be constantly and continuously supplied because of the rapid turnover and needs of red blood cells, which live only about four months. It is also needed for the growth and repair of our intestinal bacteria and the cells that line the intestinal walls.[6]

Since folic acid controls the blood levels of homocysteine, a molecule that probably causes heart disease, strokes may also be prevented if folic acid is ingested daily.[7] High doses of folic acid help people with restless leg syndrome,[8] common in those with malabsorption problems. Gout symptoms may be greatly diminished by folic acid because it inhibits the enzyme responsible for the synthesis of uric acid.[9]

Oral contraceptives contain substances that bind with folic acid, effectively removing a great deal of this vitamin from the body, making the need for folic acid urgent for these women. Another tragic and potentially alarming situation is that folic acid deficiencies have been found in two out of three geriatic patients in U.S. psychiatric wards.[10] The Food and Drug Administration believes that bread and grain products should be fortified with folic acid to prevent such deficiencies and also possibly reduce the incidence of cleft lip birth defects.[11]

Vitamin B_{12} – Energy Provider on the Cellular Assembly Line

Vitamin B_{12} (or cobalamine) is a unique cobalt-containing coenzyme that catalyzes a variety of chemical reactions. Especially intriguing to nutritionists is how this B_{12} molecule moves clusters of carbon atoms from one molecule to another. This vitamin is especially valuable on the cellular assembly line where its skills are employed to help create red blood cells and thus reverse pernicious anemia. Only certain microorganisms such as AFA are able to build this vitamin. As reported in the journal *Science*,[12] biochemists, biologists, and geneticists alike have recently come together to share ideas about how this complex synthesis might take place.

Store-bought vitamin and mineral supplements are only barely absorbed into the body no matter how much are ingested.

Although it is mostly found in animal tissue, AFA actually contains more of this vitamin than any other food source. Because vitamin B_{12} works synergistically with so many other components of AFA, it is sometimes called the "rejuvenator and energizer" vitamin. Vast numbers of physicians use vitamin B_{12} injections to "energize and revitalize" their patients. This vitamin can be obtained more naturally from AFA algae, the richest known source, as well as from fish, meat, and dairy products. Vegetarians (especially children) need to watch out for this deficiency because there isn't any vitamin B_{12} in the plant kingdom. AFA ingestion is one way the vegetarian can never again have to worry about this common problem.

Vitamin B_{12} is used in a wide variety of enzyme systems. As a result, it can be used to synthesize hemoglobin and keep blood oxygenated. It can increase energy and even repair the nervous system. Nervousness is often a sign of B_{12} deficiency and usually the first symptom to disappear after a week or so of ingesting AFA. A January 1992 study[13] reports that people over the age of 60 may develop severe neurological symptoms if they are not getting enough of vitamin B_{12}. It has long been known that vitamin B_{12} levels are always low in Alzheimer's patients.[14] If these low levels are allowed to continue, there may be some irreversible damage. The high concentration of this vitamin in AFA may turn out to effectively slow down some of the effects of Alzheimer's disease. In general, with an increase in vitamin B_{12} there is often a dramatic improvement in the mental aspects of the elderly, who seem to show signs of improved memory with less mental fatigue and disorientation.

The all-natural vitamins and minerals in AFA are 100 percent chelated so that they are all absorbed. Like drinking water from a glass, none is wasted or forced through a digestive system unable and unwilling to assimilate them.

PLEASE NOTE: some algologists claim that – with the notable exception of *aphanizomenon flos-aquae* – the vitamin B_{12} in some species of blue-green algae like spirulina contain B_{12} analogues that actually block, not activate, the natural receptor sites for this vitamin.

Vitamin C (Ascorbic Acid) – Synergistic Immune Booster

Vitamin C, or ascorbic acid, was once called the "anti-scurvy" vitamin because eighteenth century British doctors used limes, which are sources of vitamin C, to treat sailors who had a wide range of symptoms such as bleeding gums and wasting muscles. Today, many of the elderly, alcoholics, and cancer patients all over the world suffer from these and other scurvy symptoms because of a misdiagnosed vitamin C deficiency.

Interestingly, vitamin C is highly concentrated in the white blood cells of our immune system. Unfortunately, it can be quickly depleted, especially when these white blood cells have to maintain an on-going defense against tenacious invaders. It is as if vitamin C is a kind of quickly dissipating fuel for white blood cells that have been called into action by the immune system. For this reason, this vitamin must be replenished daily.

Vitamin C is actually an immuno-stimulating coenzyme. More research is needed to understand exactly how it boosts the immune system by activating white blood cell production. We already know immunity against the common cold is boosted so that the duration of illness is often reduced by roughly 30 to 50 percent, according to some studies. Although algae in general are not significantly high in this immune-enhancing vitamin, there is a powerful synergistic effect from the wide variety of AFA's immune-enhancing compounds. AFA algae has about five times the vitamin C content of chlorella and spirulina algae.[15]

Vitamin C helps to lower the amount of cholesterol overly imbedded in some cell membranes[16] and increases the strength of artery walls.[17] It seems to speed up wound healing, deter gum bleeding, maintain good vision, protect against smoking, help towards preventing diabetes, and even is useful in treating psychiatric disorders. Vitamin C content is low in the white blood cells of asthmatics.[18] As vitamin C ingestion increases, fatty acid metabolism improves[19] and bronchial constriction decreases to a more comfortable level. Even the rate of stomach, cervix, and esophagus cancer is greatly diminished with an increase in vitamin C consumption.

Vitamin C also adds alcohol groups to tryptophan molecules so that serotonin, an important neurochemical, can be formed. Tyrosine, another amino acid, needs vitamin C to be converted into dopamine and adrenalin, two important brain and stress hormones. Vitamin C also helps to add alcohol groups to other amino acids such as proline and lysine. This is good for the synthesis of connective skin tissue (collagen protein) and noticeably enhances the health of the skin.

Vitamin E – The Nerve Protector

Vitamin E (tocopherol) is a yellow oil, which, as an antioxidant, protects the cell membrane of AFA from free radical oxidative damage that originates both outside of and within the cell. Vitamin E is sometimes called the "nerve healing" vitamin and is essential to animals for its important antioxidant qualities. It is widely distributed in nature in the form of vegetable oils, whole grains, and AFA. Because of the protective presence of vitamin E, fatty acids are able to avoid free radical damage and help maintain the fluidity of the cell membrane.

It's likely that Vitamin E enhances the immune system by acting as a protecting antioxidant that strengthens cell membranes and keeps invading viruses from penetrating into the cell. Even red blood cells appear more healthy and have prolonged life, greater than the average lifespan of four months, due to vitamin E. An in-depth study given in the April 1994 issue of the *New England Journal of Medicine*[20] reports that there was even a much lower incidence of prostate cancer in male smokers who took vitamin E. British scientists have found that an increase in vitamin E will actually slow down or even reverse many neurological disorders caused by a vitamin E deficiency. This vitamin has also been shown to protect against some of the negative effects associated with radiation exposure and even seems to be generally protective against air pollution and other environmental toxins. Unfortunately, more than half of any ingested vitamin E is summarily excreted.

AFA's plentiful betacarotene is fat soluble. This why it attaches so well to the fatty acid portion of the chlorophyll-containing sections of the inner regions of the cell membrane. Dangerous fats—called LDL, or "bad" cholesterol—roaming throughout our bloodstream also attract fat-soluble betacarotene. This is profoundly beneficial to cardiovascular health. The betacarotene molecules wrap around and shield the bad cholesterol from the free radical distortion.

Betacarotene – Triumphant Free Radical Shield

Betacarotene, a precursor to vitamin A, protects the cell membranes of humans and algae alike by serving as a powerful antioxidant and free radical shield. Betacarotene may be obtained from carrots, sweet potatoes, leafy green vegetables, and certain species of algae. *AFA algae has one of the highest concentrations of betacarotene in any known food.* It is the main ingredient in the biosynthesis of vitamin A (retinol) and is therefore essential for efficient day and especially night vision. Betacarotene offers powerful anticancer, antiviral, antiaging, and antioxidant properties without the toxic affects of too much fat-soluble vitamin A.

Betacarotene even helps maintain the health of the eye covering (cornea) and stimulates the growth of new skin cells. Vitamin A and zinc work together synergistically with betacarotene to greatly improve the integrity of our stomach lining and thereby reduce the incidence of ulcers.[21] As a team, betacarotene, vitamin A, and zinc are also able to stabilize cell membranes, repair tissues, synthesize collagen, and even decrease wound healing time, especially after surgery. It is no surprise that algae has been used on battlefields for centuries to aid wounded soldiers.

There have been other interesting claims about how betacarotene most definitely boosts and enhances the entire immune system by stimulating thymus-cell production of white blood cells. Even the cell lining of both the digestive and respiratory tracts become more resistant to infection. In other studies, the number and strength of thymus-derived immune cells have been shown to be dramatically increased so that a variety of other infections can be fought off more easily. Actually, all members of the carotene family, like alpha carotene, are valuable immunostimulators. They all seem to enhance the immune system by reducing damage to the thymus gland and increasing the number and activity of white blood cells.[22]

The National Institutes of Health also confirm that betacarotene is good for the immune system and does help to prevent cancer. Large population studies of cigarette smokers have shown a decrease of lung cancer incidence with an increase in betacarotene consumption. Specifically, malignant cell growth appears somewhat suppressed by betacarotene.

The high concentration of betacarotene in AFA also supplies the vitamin A needed for the integrity of our respiratory tract lining.[23] Because vitamin A is also needed by the cells of our intestinal wall to increase mucin secretion, a healthy mucous barrier in our intestines is maintained when betacarotene-rich foods are consumed on a daily basis. A recent Finnish study found that betacarotene is best able to fight cancer when ingested in the form of food, and not as well in the form of supplements.[24]

Too much vitamin E can actually interfere with the absorption and utilization of betacarotene. AFA algae does have vitamin E (2 percent RDA), but at an amount that complements rather than interferes with betacarotene.

Biotin – Say "Goodbye" to Bad Hair Days!

Biotin is a water-soluble B-vitamin coenzyme which is important for a variety of specialized enzymes. It is able to enzymatically change neutral molecules into fatty acids and branched-chain amino acids.

Biotin is interesting because it incorporates sulfur into its molecular structure. Sulfur deficiency directly affects the health of the skin and hair. Interestingly enough, this vitamin is also produced by our own friendly intestinal bacteria.

How much biotin is needed by humans? The amount is so low (measured in micrograms) and has not as yet been established. Some claim that in larger amounts biotin may reduce baldness and graying, produce healthy-looking hair, and even tame those pesky cowlicks. It has been called the "good hair day" vitamin.

Biotin seems to reduce glucose levels in the blood and to lower the need for insulin injections for diabetics.[25] It may even improve athletic performance by helping to metabolize branched-chain amino acids.

Choline

Choline is a water-soluble B vitamin. It is one of the components of AFA's cell membrane and is essential for maintaining cellular fluidity. Choline is also needed in the biosynthesis of acetylcholine, an important neurotransmitter, and plays a role in moods and emotions. Choline is available in green leafy vegetables and AFA algae.

METABOLICALLY, VITAMINS ARE THE KEY to proper enzyme function. They animate enzymes on the cellular assembly line, allowing for life-sustaining cycles of energy and molecule transformations to take place. Equally essential to enzyme function – be it digestive or otherwise – are the mineral cofactors so plentiful within AFA. We will now take a look at this valuable and synergistic team.

Two out of three people in the United States are deficient in two or three vitamins (and at least that number of minerals) needed to activate enzymes that catalyze food into energy.

CHAPTER SIX

The Abundance of Minerals: A Treasure Trove in AFA

TODAY, IN THIS MODERN WORLD OF CHEMICALLY FERTILIZED and overworked topsoil, mineral deficiency is even more likely to occur than vitamin deficiency. The soil conditions within most farmland areas simply do not have the mineral content to produce the quality grains and vegetables that ordinarily would keep our immune systems strong and our enzyme systems working efficiently around the clock. Even if you eat a theoretically balanced diet you are still at risk of such mineral deficiencies. The mineral-rich content of Upper Klamath Lake is ideal for maintaining AFA's mineral and enzyme systems at its highest possible level.

Some minerals are the very foundation for the electrical nature of life, whereas others play out their role at the very core of every enzymatic action. Bulk minerals in the form of sodium, calcium, and potassium ions help to create the electrical impulses of muscle and heart movement as well as the electrical aspects of neurochemistry. Because some minerals are required in high concentrations, they are called "bulk" minerals, or electrolytes. Potassium is a bulk mineral that AFA also uses to bioactivate some of its protein-building enzymes, while calcium is more involved in building the cell wall.[1]

Trace minerals such as selenium, iron, copper, magnesium, or manganese help to hold enzymes together so that they might carry out one or another of their often little understood yet essential transformative biofunctions. For example, a 1995 study in *The Lancet*[2] describes how manganese occurs in very little amounts in the body, yet is responsible for biostimulating liver function by the activation of vital liver enzymes. One of the many functions of magnesium is to bioactivate enzymes that build cell membranes. This is typical of trace metals. Without them, important enzymes remain idle and useless, like once powerful and elaborate machines now silent and abandoned by operators who are always out to lunch.

Most people have a trace mineral deficiency and probably do not realize it. Roughly 30 percent of elderly people in the United States have several mineral deficiencies, even if they may appear to be in average health.[3] After all, the vegetables they eat are probably grown in mineral-deficient soil and therefore cannot be expected to rescue them from this ubiquitous and modern problem. Lacking mineral reserves, imminent "breakdowns" may be expected for these folks.

We now know that minerals should come from food whenever possible – as they once did! The wide variety of bulk and rare trace minerals in AFA fit this criterion. These minerals are easy to assimilate precisely because they are bonded to or chelated with the same amino acids to which they were originally assigned.

According to an essay written for the *Journal of the American Medical Association*,[4] the FDA and the National Academy of Sciences still believe it is premature to start raising levels of mineral RDA. Perhaps they are right after all. Natural sources of minerals do not need to be highly concentrated to contribute to good health – they need only to be absorbed and naturally chelated. Each of the minerals briefly described below is important individually. All of them are in AFA algae. Together, they form a *VITAL SUPER-SYNERGISTIC TEAM*.

BORON is the "bone-strengthening" trace mineral that is found in soil to a widely varying extent and is obtained from apples, pears, leafy greens, and especially AFA algae. U.S. government-sponsored research shows that boron deficiency may be very common, especially for the elderly and for alcoholics. Both of these groups have increasing difficulty in absorbing these minerals from their intestines.

Since boron tends to concentrate in the hormones of the parathyroid gland, it is therefore related to calcium/magnesium retention and general bone health. Soils that have high boron concentration (such as in Israel) have a lower incidence of arthritis than soils with less (for example, in Jamaica). Osteoporosis, arthritis, and hypertension seem to be lessened with boron-rich foods. Some researchers claim that boron may even enhance mental clarity. Although artery plaque and thus high blood pressure have been lessened in the presence of boron, more medical research remains to be done.

CALCIUM is the "anti-osteoporosis" mineral that is found in AFA, dairy products, leafy green vegetables, and tofu. Calcium may also be used to lower cholesterol levels in the blood[5] and protect against cardiovascular disease by lowering blood pressure.

The concentration of calcium in AFA exceeds that in milk and most other foods. Swiss cheese contains about 10 mg per gram, whereas AFA contains considerably more, at 14 mg per gram. Calcium may possibly be helpful in treating arthritis and definitely in diminishing leg cramps caused by calcium deficiency. Calcium sometimes acts as a bedtime tranquilizer with an irritability-soothing and nerve-calming effect. This is probably because some enzyme systems are bioactivated by calcium ions.

Unfortunately, as much as 75 percent of our dietary calcium normally does not get absorbed. Intestinal absorption of calcium also diminishes with age, causing calcium to be slowly removed from bones. This results in a degenerative "porous-bone" disorder called osteoporosis. It has been estimated that in the 1990s one million hip fractures and 40,000 annual deaths were caused by a deficiency of calcium and other related minerals such as boron and magnesium. Since calcium loss in bones cannot be easily replaced, osteoporosis must be prevented by consuming calcium from mineral-rich foods. Although some dietitians recommend bone meal

. . . .

PICTURE THE HEALTHY CELL . . .

as a well-run factory with a staggering variety of assembly lines working around the clock. Enzyme molecules are workers or master craftsman, depending on their level of responsibility, and nutrients such as sugars, amino acids, and fatty acids are the raw materials with which they work. Vitamins are special tools used by special enzymes – the more skilled craftsmen – to do "detail work"; minerals are used by most other enzyme-workers to bend themselves into the various shapes and positions needed to function on the cellular assembly line. As expected in any factory, "sparks fly" from the use of tools. These sparks are the free radicals which can damage all aspects of the cell and even "wound" the enzyme-workers, preventing them from working a full (ground cattle bones) as a common absorbable source of calcium, this source is often contaminated with lead! Moderate exercise, stress reduction, and ingestion of food sources rather than mineral supplements will improve calcium absorption. The calcium present in AFA is actually easier to absorb than that in many other foods. This is because the betacarotene and amino acids in AFA algae are able to wrap around, chelate, and thus more easily transport calcium through our intestinal wall.

CHROMIUM is a trace micromineral with just 5 mg distributed throughout the entire human body. It is the "sugar-regulating" trace mineral found in most commonly in brewer's yeast, black pepper, and AFA. Chromium also bioactivates a variety of enzymes. As an enzyme cofactor, it activates insulin and thus controls glucose levels. This is beneficial for diabetics, hypoglycemics, and those with high cholesterol. Chromium also stimulates fatty acid and cholesterol synthesis in the liver. To some extent, it also assists in the digestion of proteins in the intestine.

Keep in mind, however, that refined sugars should be avoided. They will only cause the body to excrete the small amount of chromium it may have absorbed. There is some evidence that refined white flour removes even more chromium from the blood. To make matters worse, chromium is relatively difficult to absorb in the first place. For example, inorganic chromium chloride when clinically admin-

. . . .

day. Safety goggles and shields – antioxidants – are needed for continuous protection. Just as the factory assembly line comes to a standstill when workers are deprived of their tools, so too are enzymes dysfunctional when deprived of their vitamins and minerals. The wrong mineral – such as the environmental toxins lead and cadmium – can sometimes distort an enzyme into a shape that impairs the work of the entire cellular factory. And just as the assembly line in a factory stops when a raw material is no longer supplied, our body cannot function properly when certain amino acids are deficient. Dietary amino acids must all be available and ingested simultaneously or be wasted. AFA ALGAE HAS ALL AMINO ACIDS IN JUST THE RIGHT PROPORTIONS.

istered is only 5 percent absorbed.[6] The rest is excreted. Fortunately, absorption may be raised to 20 percent if the chromium is bonded to amino acids, as it is in AFA. This kind of good chromium absorption may help to prevent adult-onset diabetes and most hypoglycemic symptoms.

Unfortunately, our mineral-poor topsoil, along with our over-refined foods, has led to an alarming depletion of chromium in the U.S. diet. The use of high-fructose corn syrup as a cheap sugar substitute has also increased dramatically in the last 15 years.[7] This effectively reduces chromium levels in the general population. As a result, is it any wonder that the rates of diabetes, heart disease, and atherosclerosis are much higher in the United States than in other parts of the world? Diabetes, for example, has steadily increased by *600 percent* since the 1940s.

Chromium is an essential part of a relatively large glucose-regulating molecule called the "glucose tolerance factor" (GTF). The exact structure of this molecule is unknown. As part of GTF, chromium allows insulin to bond to receptor sites on glucose-hungry cells so that glucose can pass more easily from the blood into the cell. Excessive glucose circulating in the blood may lead to diabetes. To make matters worse, excessive glucose burns and damages artery walls. Aging, pregnancy, and overexercise are factors that decrease chromium absorption and increase its excretion.

COBALT is an essential part of vitamin B_{12} (cobalamin) and assists in the complex biosynthesis of red blood cells as well the maintenance and repair of nerve tissue. Cobalt is found in ocean fish, red meat, liver, and AFA algae. The high concentration of cobalt in AFA is good news for vegetarians since the soil content of cobalt is normally very low, reducing further the already low amount of this trace metal in plants.

COPPER is an essential trace mineral found in liver, shellfish, and AFA. Copper is an important micronutrient in many enzyme systems. With the help of iron, it is important in the biosynthesis of hemoglobin. Once copper is absorbed, it is escorted through the bloodstream by specialized carrier proteins to the liver where it is biosynthesized into an antioxidant blood protein (ceruloplasmin), which absorbs free radicals. This helps to prevent fatty acids from turning rancid and therefore it helps to keep cell membranes from becoming less flexible. Copper is also part of the antioxidant enzyme superoxide dismutase, which may have an antiarthritic effect because it keeps the synovial joints fluid.

Zinc and copper work well together. But since both minerals compete for the same binding sites, it is important that their ratio be "balanced" so that each mineral can perform its role effectively.[8] Few nutritionists and physicians agree on what that balanced ratio should be. The ratio of zinc to copper in AFA is 4.7 to 1, a reportedly good ratio for the absorption of both. When there is an optimal ratio of zinc to copper, cardiovascular disease is diminished to some extent because cholesterol levels are kept relatively low. However, too much copper from polluted water may lead to symptoms such as mild senility and even schizophrenia.

Why does AFA – small and simple as it seems to be – contain more micronutrients than any other known food? It is partly because AFA utilizes them to create the best enzyme system possible to blend together the four most basic ingredients available – earth, water, air, and sunlight. It is also because AFA cells are about 20 to 30 times smaller than the cells within the food we usually eat. Because of this, AFA contains 20 to 30 times the membrane surface area and therefore that many times more micronutrients!

FLUORINE is a poisonous diatomic gas. However, in the negatively charged fluoride ion form, it is believed to be potent against tooth decay and possibly to help slow down osteoporosis. A controversial study of 40,000 children who drink fluoridated water showed a 36 percent drop in tooth decay, and 50 percent of the children did not have any cavities at all.

Apparently, fluoride in our water (or our algae) is also protective against osteoporosis because it biostimulates new and stronger bone growth. People who live in areas with fluoridated water have a dramatically lower incidence of collapsed vertebrae phenomenon. There is a considerable decrease in soft tissue calcification for those who drink fluoridated water. Their arteries actually remain more flexible. In the United States, we get most of our fluoride ions daily from drinking water. However, many of us use bottled water for a variety of good reasons and need to depend on other fluoride ion sources such as seafood and AFA algae. *

GERMANIUM was thought, back in 1987, to be the new miracle micromineral. There is some evidence that organic germanium compounds do stimulate the immune system and show anticancer activity. There has even been some success in controlling the virus of the chronic Epstein-Barr disease. Over a long period, the low concentration of germanium in AFA may yield some of these positive healthful benefits.

IODINE is a trace mineral needed by the thyroid gland of the endocrine system to biosynthesize a variety of hormones such as thyroxine, which controls energy metabolism and the rate at which betacarotene is converted to vitamin A. Goiter (swelling of the thyroid) and cretinism (mental retardation) are well-known diseases that result from iodine deficiency.

Iodine must be supplied daily because our bodies cannot store it. The best sources of daily dietary iodine are kelp, seaweed, ocean fish, and AFA. Irrigation water as a successful method of supplying iodine to severely deficient populations in Xinjiang, China, was reported in *The Lancet*.[9] Unfortunately, our infamous fast foods are actually too high in iodine ions because of excessive iodized salt. Such high dosages of iodine from fast foods will often cause acne problems.

Iodine offers protection against the toxic effects of radioactive substances in the environment. Apparently, the presence of nonradioactive iodine in the thyroid keeps radioactive iodine from accumulating there.

IRON is the "rosy cheek" trace mineral that is absolutely essential for human survival. Hemoglobin molecules, which give blood its red color, contain four iron ions, each of which are used to transport oxygen (O_2) molecules to every cell of our body. Iron has played a successful role in exotic enzyme systems which have been used by ancient microorganisms such as blue-green algae for 3-4 billion years.

Iron is available in red meats, salmon, brewer's yeast, kale and other leafy greens, and AFA. The absorption of iron takes place across the intestinal walls over a period of roughly three hours from ingestion. Vitamin C increases absorption, as do small amounts of copper, cobalt, and manganese, all of which are present in AFA. To maintain high iron absorption, the use of antacids should be minimized. Since such digestive aids tend to reduce stomach acidity, iron compounds are not able to break apart for absorption purposes. Also, excessive calcium, aspirin, and tetracycline should be avoided because they all compete with iron for intestinal absorption. The low-income elderly in the United States are at great risk for having iron deficiency.[10] There is mounting statistical evidence that roughly 40 percent of the entire U.S. population have iron levels below what is considered healthy.[11]

Fatigue, irritability, depression, and low enthusiasm are all typically symptomatic of iron-deficient anemia. The cure is usually the assimilation of more iron. This is due to the bioactivation of iron-containing enzymes, which are essential to DNA synthesis.

The immune system is also biostimulated with increased iron assimilation. Even some of our more specialized white blood cells need iron to bioactivate exotic lactoferrin proteins. These proteins help to generate beneficial oxygen radicals, which are used as "bullets" to destroy invading bacteria.

When it comes to children, any mineral nutrient deficiency can impair the function of the brain and lead to a variety of learning disorders. Because of serious deficiencies in our farming soil, iron ranks as the mineral most lacking in American children today. As the availability of iron decreases in the body, so too does a child's attentiveness. It has been shown by biochemists and research physicians alike that even subtle negative fluctuations of nutritional minerals can substantially impact learning abilities in children.[12] A good day in school for a child is directly related to the minerals they are getting on a daily basis. Rather than guess at their daily mineral intake, the bioavailable minerals in AFA algae are strongly suggested.

MAGNESIUM is important as an "antistress" mineral and plays a role in more than 325 different enzyme systems. Magnesium helps control almost every one of our vital biochemical processes. For example, magnesium is part of an important enzyme mechanism that synthesizes muscle protein and repairs worn out collagen. It even seems to relax the bronchial muscles in asthmatics.[13]

Getting enough magnesium contributes toward making us feel healthy and vibrant. Unfortunately, magnesium deficiency is much too common. The USDA tells us that the average American gets only 25 percent of the RDA of magnesium! Since magnesium atoms are always found in the center of green chlorophyll molecules, all leafy green vegetables are expected to be high in this valuable and relatively rare mineral. Of course, since AFA has one of the highest percents of chlorophyll of any food on the planet, it is no wonder that AFA algae are also rich in magnesium. When delivered to our bodies chelated to chlorophyll, magnesium is especially easy to assimilate.

Magnesium is important for the proper functioning of the heart and can be used to treat irregularities in heartbeat rhythm. It can also be useful in elevating high-density lipid (HDL, the "good" cholesterol) proteins as a prevention for atherosclerosis. Magnesium has even been shown to decrease the stickiness of platelets, which decreases the tendency of blood to clot and cause high blood pressure.[14]

Since hard water is high in calcium and magnesium, people in geographical areas that have hard water have a lower incidence of cardiovascular problems.

Magnesium deficiency is known to cause tiredness and irritability.[15] Its deficiency is known to increase the incidence of migraine headaches as well as symptoms of irritation and mood swings. Long-term magnesium deficiency can even lead to confusion and anxiety. Because of these factors and magnesium's effectiveness as a natural tranquilizer, this mineral has also been used in treating certain mental illnesses.[16]

People who drink milk for calcium are up against a real Catch-22. Most people are unaware of the fact that the calcium in milk actually has a lowering effect on magnesium availability. Because of this link, milk consumption can be connected with heart disease as well as, for the same reason, an increase in osteoporosis.[17] Drinking milk fortified with vitamin D may also result in lowered magnesium levels.[18] Drink milk – indeed. In addition, the phosphate ions of soft drinks will bond to and remove magnesium, and magnesium absorption is also low in the presence of sugar.

MANGANESE is a little-understood trace mineral found in small amounts in leafy greens, nuts, whole grains, and AFA, and it is important in many enzyme systems. For example, in one enzyme system it is responsible for catalyzing the biosynthesis of dopamine, an important neurotransmitter. In other systems, it assists in the formation of bones and the metabolism of glucose. Manganese even plays a crucial and protective role in the antioxidant molecule superoxide dismutase.

Unfortunately, high-tech farming depletes manganese from the soil. In addition, the process of refining wheat to white flour typically depletes 80 to 90 percent of the available manganese. To make matters worse, only 15 percent of our dietary manganese is normally assimilated. Those of us who consume meat, dairy, and soda pop have an even more difficult time absorbing this trace metal.

The manganese in AFA is relatively easy to assimilate. For billions of years blue-green algae have used their manganese in the process of photosynthesis. Since the manganese is part of an elaborate enzyme system, it is chelated and much easier to absorb. Once absorbed, manganese can be helpful in relieving fatigue and nervousness, as well as increasing production of mucopolysaccharides needed for healthy bone joints. Manganese is typically deficient in those who have rheumatoid arthritis.[19]

Thankfully, manganese can be stored for later use. A beautifully specialized spherical protein (transmangamin) is able to carefully transport any excess manganese to the liver for storage. It may then be called upon to help prevent cancer since all cancerous tumors are exceptionally low in this mysterious mineral.

MOLYBDENUM is a very rare mineral and important micronutrient. It is available in AFA algae and can bioactivate several human enzymes associated with longevity enhancement and free radical absorption.

NICKEL plays an essential role in nutrition. Present in AFA, nickel is needed for growth and reproduction.[20] Nickel was found as a cofactor inside of the first enzyme ever purified in the laboratory.

POTASSIUM coupled with a low intake of sodium, reduces the blood-vessel constricting effects of the adrenaline released during mental stress. Potassium is also important to electrolyte and acid-base balance inside human cells.

SELENIUM received some interesting attention in *Science News* in May 1995.[21] Apparently, certain viruses remain harmless in mice until their selenium levels drop to the point of impairing the immune system. When the immune system is low, the virus is able to mutate into even more dangerous forms. AFA offers a wide variety of immune-stimulating substances that protect us on a daily basis in different ways. Selenium may also increase the elastic youthfulness of the skin and be helpful in removing age spots. When ingested in the natural algae form of selenomethionine, skin cancer incidence from ultraviolet light may also be reduced. Some forms of algae found in the Great Barrier Reef of Australia contain similar amino acids which show promise as sunscreen compounds.[22] Selenium may also be helpful in improving the function of the cell membrane.

SILICON in AFA algae, and in the skin and membranes of fruits, is a useful raw material for strengthening the human skin. Ingestion of excessive silicon leads to a decrease in the biosynthesis of neurotransmitters such as serotonin and dopamine.[23]

ZINC is a part of more enzyme systems than any other mineral. Its enzymes play a key role in the repair of the single strand of DNA that controls the genetic activity of blue-green algae. AFA uses specific zinc enzymes to help build the "cobblestone" structure of its flexible cell membrane.[24] Zinc is used by humans to activate digestive enzymes that manufacture necessary stomach acids. When dietary zinc is low, so too is our stomach acid and the ability to digest food and assimilate nutrients. Zinc deficiency and low stomach acid are especially widespread in the elderly.[25] Today, even infants who are breastfed may not be getting enough zinc.

A report in *The Lancet* showed how zinc supplements can increase growth in infants even if they are breastfed. [26]

Zinc works synergistically with betacarotene to protect our cell membranes. It is one of the minerals (along with copper and manganese) that activates the antioxidant enzyme superoxide dismutase (SOD), which offers some help to rheumatoid arthritis. The antioxidant properties of some zinc enzymes even reduce plaque formation in the gums and help to prevent periodontal disease.[27]

Canker sores, which probably affect a large percent of the population, are usually blamed on stress and nutrient deficiency. Getting enough zinc will keep these painful sores from forming.

Prostate enlargement (prostatitus) affects 60 percent of men over 60. Zinc inhibits the enzyme that causes this enlargement.[28]

Heavy metal contamination of the blood leads to a variety of problems. Cadmium, for example, is known to cause hypertension and high blood pressure in laboratory animals. A daily intake of zinc may begin to reverse such cadmium-induced hypertension.[29]

WHILE MINERALS ONLY REPRESENT about one percent or less of our daily food intake, without them we cannot utilize our food, and we would quickly perish. Our very enzymatic biological machinery within every one of our cells would be silenced.

Remember – nearly all of the minerals of AFA are held tightly and deeply within the very core of hundreds of transformative enzyme systems. Whether it is bone-strengthening boron and calcium, sugar-regulating chromium, or immune-boosting zinc and copper, such naturally-chelated minerals are directly assimilated and more easily put to work within our own similar, but vastly more complex, system of enzymes. As a result, each of our cells function more smoothly so that our entire immune system – as we'll see in the next chapter – is biostimulated to a new level of vibrant and long-lasting health.

CHAPTER SEVEN

Fortressing the Immune System

THE HUMAN IMMUNE SYSTEM IS A COMPLEX AND MIRACULOUS wonder. Based upon a dynamic and profoundly cooperative arrangement of specialized organs, bone marrow, and body-guarding cells, the immune system protects us from a wide variety of continually invading foreign cellular threats such as bacteria and viruses. This same immune system is also responsible for protecting us from our own potentially cancerous cells that may traitorously turn on us at any time for reasons science cannot as yet fully explain.

As a result, we must keep our immune system strong at all times and throughout our lives. This task is especially difficult in the midst of our chemically-polluted and anxiety-ridden world. When the immune system begins to weaken, we open ourselves up to an insidious assortment of chronic infections.

Americans and Western Europeans today are at risk for a number of diseases because of their chronically depressed immune systems.[1] For example, chronic fatigue syndrome becomes aggravated only when the immune system is compromised. This is probably because the supposed culprit, the Epstein-Barr virus, which is latent in almost half the U.S. population, lies dormant only until our defenses can no longer prevent its spread. There are many more diseases waiting to compromise our health. We desperately need to reverse this trend and regain our health. But to do this properly, we really need to conceptually understand just how our immune system works and how we can fortress and boost it to new levels of protection.

Ingesting AFA algae on a daily basis is one way to help biostimulate and strengthen our immune system. Eating nutritional foods and avoiding those that are unhealthy should be another necessary and complementary daily habit. Calamitously, the average American consumes about 125 pounds of sugar per year. Many of us do not understand just how negative an impact this has on our immune system. Ingestion of large (and especially daily) quantities of sugar considerably decreases the ability of our white blood cells to destroy invading bacteria and viruses.[2] It is for this reason that we must avoid any sweets or even fruit juice during the first few days of the flu or common cold. Fortunately, AFA eaters experience a considerable and noticeable decrease in the craving of sweets, so our immune system begins to strengthen and improve right from the beginning.

Lymphatic Fluid – The Intercellular Cleanser

It is surprising to learn that about one-seventh of our body is actually filled with an intercellular healing fluid that ceaselessly flows between the spaces of our cells. After miraculously carrying away wasted and worn-out molecules, this liver-derived essential fluid enters our lymph vessels as "lymphatic fluid," traveling parallel to the arteries and veins of our circulatory system. At specific locations in the body, this lymphatic fluid then pours into various lymph nodes, where it is purified by the concerted action of a team of interesting white blood cells, described below.

T-Cells are our body-guarding, lymphatic white blood cells, which originate in the bone marrow and mature in the thymus gland. These thymus-derived lymphocytes, or simply T-cells, tirelessly travel through the lymph vessels and intercellular fluids as our valiant defenders, examining our tissues, and occasionally giving the chemical order to other white blood cells, called "natural killer" cells, to attack any cell that has been contaminated with a foreign invader virus hiding inside. The natural killer cells promptly obey their orders by blasting the impure cell with a glob of phospholipase, a zinc-dependent enzyme that attacks and dissolves the cell membrane of the invaders. This is one reason why the zinc in AFA is useful in enhancing our immune system.

Betacarotene is also crucial for enhancing and strengthening our immune system. The vast stores of carotene compounds hidden deep within the photosynthesizing thylakoids of AFA algae biostimulate our otherwise aging and shrinking thymus gland to secrete more thymosin, one of its important hormones. This hormone, along with zinc and arginine, AFA's second most concentrated essential amino acid, encourages the thymus gland to produce even greater numbers of more active and energetic T-cells.

Such energized T-cells are then better able to produce *interferon*, a molecule that binds to human cell surfaces and stimulates protein synthesis so that cell walls are strengthened and more actively protected from viral invaders. Interferon binds to cell surfaces and stimulates protein synthesis, which strengthens the cell wall and prevents viruses from entering and infecting them. Proper exercise and adequate sleep also stimulates the immune response by increasing the protection given by interferon.[3] This phenomenon, along with proper diet and AFA consumption, quite literally "fortresses" and protects our immune system.

Fish oils, certain vegetable oils, and the essential fatty acids of AFA all contribute toward an increase in the number of *suppressor* T-cells originating from the thymus gland. One noticeable and fortunate effect from these immune cells is a decrease in some all-too-common food allergy reactions. For example, dark regions under the eyes, what some physicians call "allergic shiners," tend to diminish or even disappear over time when T-cells are increased.

B-Cells are lymphocytes found within small sacs called bursae, which are located between bones and tendons. After a "helper" T-cell has identified an invad-

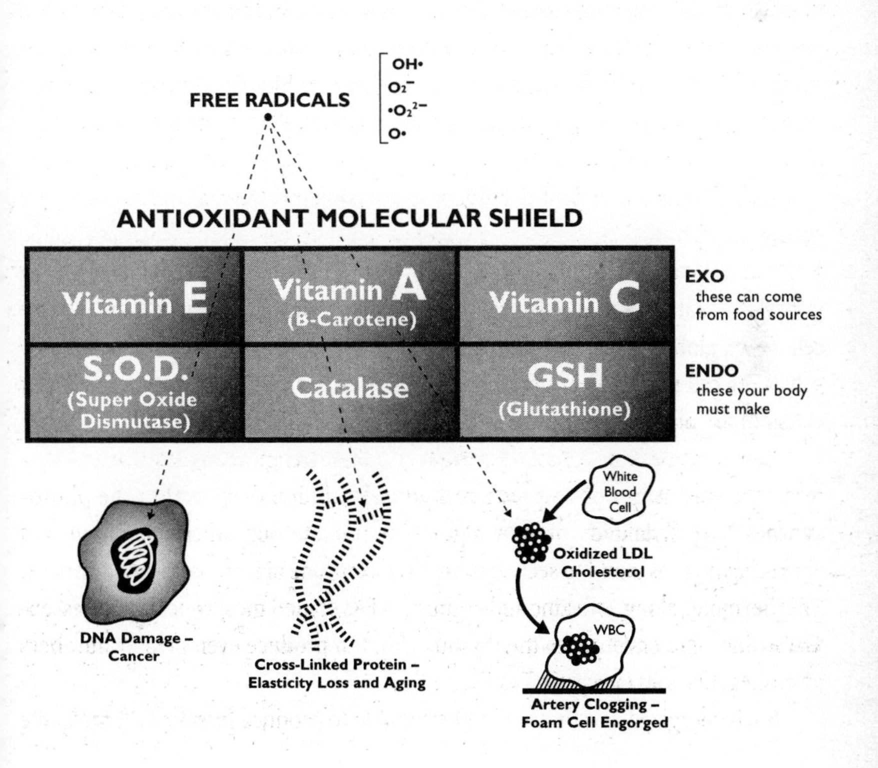

ing organism, it guides our B-cells to the invader whereby a specialized antenna-like glycoprotein on the B-cell attaches to the invader. Once attached, the B-cell divides and multiplies into larger *plasma* cells that secrete protein material called antibody molecules. These molecules then attach to the invading cell, triggering the production of another more powerful *complement* protein that eventually punctures and kills the invading cell. Amazing!

Because antibody proteins have an overall globular shape, they are also called immunoglobulins. There are five human types, characterized by their polypeptide arms that can reach out and hold two or more invading organisms. The difference between one antibody and another may sometimes be nothing more than the difference of several atoms on the tips of its receptor site arms.

"Synergistic biostimulation" is a sign of health and occurs when enzymes and vitamins begin to activate otherwise hibernating enzymes. When a group of such enzymes begins to work together toward the common goal of cellular integrity, something wonderful happens. The whole becomes greater than the enzymatic sum of its parts. The symphony of health begins.

Immunoglobulin G (IgG) is the most common antibody. In a study some subjects were able to actually boost their immune systems and increase the immunoglobulin A (IgA) concentration in their saliva simply by favorably viewing humorous videotapes.[4] Positive emotional states do seem to have a real and verifiable uplifting effect on the immune system. Conversely, it is also an established fact that psychological stress factors can subtly damage our immune system.[5] As mental stress increases, some immunoglobulins unfortunately are found to decrease and thus impair the immune system.

Neutrophil cells mainly engulf and destroy bacteria by phagocytosis, which means they surround and then digest the bacteria with secreted enzymes. Neutrophil cells represent about two-thirds of the total circulating white blood cells. All white blood cells, like other cells of the body, require vitamins and minerals to properly function. However, high intake of sugar products is known to impair their proper function.[6]

Monocyte cells are large cells that engulf cellular debris left over after an infection. Macrophage cells are monocytes that engulf bacteria and cellular debris that originate in the gastrointestinal tract. Basophil cells release histamine so as to break apart antigen-antibody complexes. Allergic reactions sometimes follow.

Betacarotene – Our Heroic Friend

AFA ALGAE RANKS AMONG THE HIGHEST IN BETACAROTENE concentration of any known food. As a rich source of betacarotene, AFA is able to boost our immune system capabilities by increasing the number of thymus gland lymphocyte cells and enhancing the production of interferon. Betacarotene and the other carotenoid compounds of AFA biostimulate our immune system by increasing the number, activity, and circulation of antiviral thymus helper cells.[7] And once betacarotene has been biochemically transformed into vitamin A, white blood cells may use it to increase their abilities to kill invading viruses. This vitamin A precursor is also known to increase B-cell activity, thus increasing antibody production, especially IgA, when necessary.

Those of us who keep getting colds should take a good look at how much stress is in our lives. Stress directly affects the adrenal glands and tends to release a family of interrelated molecules that cause the thymus gland to shrink in size and significantly diminish its activity. Aging also causes the thymus gland to shrink and produce less thymic hormones such as thymosin. Betacarotene enhances and protects the thymus gland from these effects of stress and aging. Betacarotene also helps to promote healthy lung and digestive tract linings. As a result, respiratory and digestive tract infections are reportedly diminished because pathogens cannot penetrate the stronger cell membranes.

One bothering indicator of a depressed immune system is recurring attacks of boils. Many naturopathic doctors recommend taking zinc and betacarotene supplements as a preventative.[8] Since these immune stimulating substances are generously and naturally supplied by AFA, it should be no surprise that such boils either would no longer occur, or if they did, would do so with less severity.[9]

Betacarotene also helps to slowly push back a high population of *Candida*, a yeast-like pathogenic microorganism, by increasing the activity of neutrophils which reduce their numbers. This is especially helpful combined with a good probiotics approach (see Chapter 13).

There are definite advantages to utilizing a "flash-frozen freeze-dried" technique when harvesting a bloom of AFA algae. Sun-drying algae, for example, will diminish its concentration by about 30 percent of methionine and a devastating 50 percent of betacarotene.[10]

Improving Cell Membrane Fluidity

Essential fatty acids such as gamma-linolenic acid located within AFA's flexible cell membrane help to enhance the fluidity of the cell membranes of all white blood cells. This leads to their greater mobility and phagocytic abilities, both of which are connected to ridding our body of foreign invaders and thus enhancing the immune system. Eating AFA should, over time, improve the quality of the fatty acids used in keeping cell membranes fluid. This is explored further in Chapter 8.

Immunity and Free Radical Protection

Complex energy-producing reactions continuously take place within all of our cells. There, deep inside energy factories called mitochondria, a vast array of nutrient molecules are efficiently and enzymatically oxidized to provide the energy required to maintain the life processes of the cell. During this complex biomechanical process, simple yet incomplete molecular fragments called free radicals are produced. Because each fragment contains an odd number of electrons, it is free and radically compelled to rob a single electron from any biomolecule it accidentally crashes into, oxidizing it and thus altering its shape and biochemical function. This is often called oxidative stress. Antioxidants such as betacarotene are needed to shield us from these dastardly arrows within.

The *Journal of the American Medical Association* concluded in 1993 that these free radicals really do cause substantial tissue damage and can dramatically injure any vital organ by altering the structure and function of the fatty acids found in all cell membranes.[11] The following year, the prestigious journal *The Lancet* presented overwhelming evidence that free radicals cause a variety of diseases.[12] Environmental radiation and physiological processes in the body cause free radicals to form, which in turn damage cells and the delicate DNA within. Enzymes and other antioxidants in our body work around the clock to repair this damage.

The presence of antioxidants with free radical–quenching abilities is especially beneficial to the immune system. Antioxidants are needed to protect the cell membranes of white blood cells to enable them to travel through the lymph system and function more efficiently. AFA's betacarotene and selenium-dependent enzyme, glutathione peroxidase, specifically protects our immune cells against the damaging effects of peroxide free radicals. Such antioxidant systems are especially beneficial in preventing damage to the thymus gland and thus the entire immune system. The sulfur-containing amino acids of AFA, methionine, and cystine also have antioxidant abilities, as do the selenium ions to which they are often chelated.

There are other free radicals that are also dangerous to the health of the cell membrane and the vitality of the immune system. The superoxide free radical may be stopped before it does its damage by a powerful antioxidant enzyme system within AFA algae, aptly named superoxide dismutase. SOD is an antioxidant that may be helpful in preventing further complications of Crohn's disease. This is presently being researched at the Institute of Biophysical Chemistry in Paris. SOD may be bioactivated by one of three minerals: zinc, copper, or manganese. Fortunately, AFA contains all three minerals.

The betacarotene in AFA is probably one of the most powerful natural antioxidants known today. It May well be the single most important factor towards enhancing the immune system (by protecting the thymus gland) and thus increasing our life span. However, the potential healing properties of betacarotene are maximized ONLY *when other precious carotenoids are also present. AFA contains dozens of carotenoids, all able to neutralize the deleterious effects of dangerous free radicals.*

There are other AFA minerals with immune-boosting potential as well:

- *Chromium is needed to help insulin molecules inject blood glucose across the cell membrane into waiting and hungry cells. This same property may also help lymphocytes function more efficiently as members of the immune system.*

- *Copper deficiency sometimes leads to a decrease in the amount of antibodies produced after a viral or bacterial invasion. The trace amounts of copper present in AFA, along with its larger amounts of zinc, are helpful in maintaining B-lymphocyte function and antibody production.*
- *Germanium has been shown to increase macrophage activity, which protects the body against infection, in laboratory animals. Germanium also may possibly increase interferon production and possess anticancer advantages. In addition, germanium has been successful in controlling the virus of chronic Epstein-Barr disease. The small amount of germanium in AFA may be enough to contribute to a variety of such positive healthful benefits.*
- *Iron has been shown to enhance the activity of T-cells. Apparently an iron-dependent enzyme is involved in the DNA synthesis of these immune cells. Low iron simply means low T-cell count. Also, neutrophils require iron to produce a rare protein (lactoferrin), which is able to destroy invading bacteria by generating free radical "bullets" called reactive oxygen species (ROS). ROS and lactoferrin are currently being studied and used experimentally in immune deficiency diseases such as HIV.*
- *One can determine if there is an iron deficiency (and possible anemia or a poor immune system) by measuring the amount of ferritin, an iron storage protein in the blood. Oftentimes, with the help of a microscope, one can diagnose iron deficiency by counting the number of small-sized red blood cells.*[12]
- *Zinc deficiency has been often shown to reduce antibody production and diminish white blood cell activity. This may be particularly significant for the older population and lead to possible auto-immune disorders. AFA ingestion is strongly suggested on a twice daily basis.*

Most of the biochemical components in AFA which help to enhance our immune system are present as micronutrients. Any one of them alone may not be enough to increase the production of T-cells, natural killer cells, or macrophages. But all of them together have a synergistic effect that is probably much greater than a large amount of any one component alone. They all seem to work as a team, each enhancing the immune system in its own way. For example, the zinc and beta-carotene of AFA function well together to boost the immune system by helping to manufacture thyroid hormones and improve our resistance to antigen attack.

Zinc is also important in boosting the immunity of individual cells[14] and the immune activity of natural killer cells. Keep in mind that zinc is able to do all this because it plays a key functional role in hundreds of crucial enzyme systems that watch over the entire immune system. Zinc is able to hold and bend the protein to which it is attached in just the right shape so as to properly activate it as an enzyme system. Without zinc, the panoply of enzyme functions would come to a grinding halt and the immune system would quickly and noticeably weaken.

ARGININE is a conditionally essential amino acid required by children and stress-affected people who cannot biosynthesize it. Arginine biostimulates the thymus gland and thus enhances the entire immune system by producing more T-lymphocyte and natural killer cells. Arginine increases the effectiveness of these cells in fighting infection and inhibiting tumor growth. Fifty years of vast research evidence attest to arginine's antitumor effect.

There is no doubt that the arginine in AFA has immune-stimulating capabilities. By stimulating the thymus gland, there is a significant increase in lymphocyte production and activity. As a direct result, diseased cells will be found sooner and destroyed more efficiently. An important and recent study by Japanese medical researchers showed how arginine actually increased the activity and fervor of natural killer cells in patients by several hundred percent![15]

Cell Wall Compounds Are Also Immuno-stimulating

The glycogen-related carbohydrate compounds, as well as the glycolipoprotein materials of AFA's soft cell wall are complex polysaccharides that even in small quantities, enhance the immune system. These cell-wall substances are somewhat structurally similar to the complex sugars found in echinacea, a well-known immuno-stimulating herbal remedy. Such polysaccharide compounds are believed to enhance the phagocytic "engulfing" effects of macrophage white blood cells against tumor cells. These compounds also enhance the ability of macrophage cells to destroy invaders using free radicals to attack them.[16]

IN CHAPTER EIGHT WE TAKE A DEEPER LOOK at the synergistic effect of AFA's antioxidant components and how they can affect the aging process by protecting us from the free radicals inside.

CHAPTER EIGHT

Antioxidants, Aging & Longevity

OUR FASCINATION WITH THE VARIOUS LEGENDS SURROUNDING our quest for the so-called "fountain of youth" or those mythological "magic potions" that whisper possibilities of immortality are probably based upon our wanting to simply avoid the detrimental changes and often deteriorating effects of growing old.

Although many of us are willing to accept the majority of life's irreversible processes associated with old age, few of us realize just how many choices are now within our reach to improve the quality of our lives.

High blood pressure and hypertension are being found with greater frequency in developing countries because of their greater dependence on modern diets of processed food. People from the unpolluted countryside of Africa or China, on the other hand, simply do not show signs of increasing blood pressure with increasing age.[1]

Antioxidants as Free Radical "Molecular Shields"

There is now a vast amount of scientific research that clearly demonstrates that the life span of any cell is directly related to the amount of protective substances or internal "molecular shields" contained within it. As discussed in Chapter 7, these molecules are called antioxidants because they protect us from the ravaging effects of dangerous forms of oxygen called free radicals.

Such free radicals are continually and naturally generated inside every cell during the many steps involved in cellular metabolism. In the form of peroxides and superoxides, free radicals can burn or oxidize portions of the cell membrane or destroy "rungs" of the cell's DNA "ladder." Dysfunctional cell membranes and altered DNA are both at the root of the aging process, resulting in a variety of familiar effects such as atherosclerosis, cataracts, cancer, Alzheimer's disease, and many immune deficiencies.

Free radicals are also found outside of us as well, increasing our needs for as much antioxidant protection as possible. Barbecued and charbroiled foods, as well as alcohol and coffee, contain much of these pollutants, as does tobacco smoke, air pollution, UV radiation, pesticides, and other industrial chemicals. Sometimes the mere act of breathing can produce free radicals that promote aging and increase the risk of heart attack and cancer.[2] The wide spectrum of antioxidant enzymes and vitamins in *aphanizomenon flos-aquae* can be very beneficial in protecting us from all types of free radicals.

All living cells require a variety of antioxidant molecules carefully placed so that each part of the cell is protected and its vital metabolic processes guaranteed to function for as long as possible. We humans live roughly twice as long as chimpanzees mainly because of our higher concentration and wider variety of internal and protective antioxidants.

All antioxidants from any species share one attribute. Whether they wait within a cell membrane, roam in the cytoplasm of the cell, or wander in the bloodstream of any animal, antioxidants all have the ability to provide free radicals with the one electron they each are so dangerously missing. Once done, the free radical "bomb" is defused, and the molecular danger is temporarily over.

> According to a study reported in *The Lancet*,[3] daily intake of betacarotene leads to a lower risk of having a heart attack. The 683 heart attack patients in this study were significantly protected because of diminished low-density lipoproteins (LDLs). Oxidized LDLs are believed to be responsible for the build-up of artery-clogging plaque, and thus bear the label "bad cholesterol."

Cell Membrane Protection Preserves Flexibility

Imagine, for example, just how important endogenous antioxidants are to the life of blue-green algae. By diminishing the deteriorating free radical effects on the cell membrane lipids, the fluidity and flexibility of that membrane is preserved. The cell lives longer.

When we ingest AFA algae, we are inviting into every cell of our bodies the inner protection of a wide variety of antioxidants. Some help to protect the membranes of our white blood cells or thymus gland cells, thus improving and strengthening the immune system. Others protect the DNA of our liver and pancreas cells to insure that they correctly biosynthesize the right enzymes and proteins so the entire body functions smoothly.

Free radical damage may also lead to the peroxidation and subsequent distortion of the fatty acids that compose our cell membranes. Such distortions may even change receptor sites on the membrane so that incoming messenger hormones will not be able to relay important information to the cell. Hence, the cell becomes dysfunctional.

In general, antioxidants keep our cells youthful by protecting them from the free radical fires within. Some reported benefits of ingesting antioxidants daily are:

- *longer preservation of skin, muscle, and connective tissue elasticity*
- *regeneration and strengthening of heart muscles*
- *less cardiovascular disease*
- *stronger immune system*

In addition, by reducing free radical damage to our DNA molecules, we minimize the biosynthesis of imperfect enzymes and the cellular confusion (possibly cancer) that may result.

There are other types of even more exotic free radicals called reactive oxygen metabolites (ROMs). These are produced in the body in response to the inflammation caused by foreign organisms and may lead to a build-up of rancid fats and dysfunctionally oxidized proteins. Too many free radical ROMs in the system often leads to chronic inflammation and eventually bowel diseases such as colitis. Such ROMs have been treated with synthetic antioxidants such as mesalazine, since antioxidant supplementation brings down inflammation. However, betacarotene is even more effective. AFA has one of the highest known concentrations of betacarotene along with a wide variety of other antioxidants that are available for reliable free radical protection.

Glutathione Peroxidase – The Ancient Antioxidant Shield

This tripeptide antioxidant molecule is an ancient one. It may have been the first molecular shield to protect primitive forms of blue-green algae from the very oxygen they were just beginning to produce billions of years ago. Actually, as the word peroxidase implies, the purpose of this molecule is to destroy any peroxides that attach to the fatty acids of the cell membrane. This is one way that the flexibility of the cell membrane is maintained. All living cells, to varying extents, universally have this tripeptide antioxidant near or attached to their outer and inner cell membranes.

Low levels of this antioxidizing enzyme can be raised to normal with the help of selenium and vitamin E in AFA. When we ingest AFA, we also benefit from the three amino acids that compose GSH: cystine, glycine, and glutamic acid.

Cystine is particularly interesting because it also contains a sulfur-hydrogen (SH) pair of atoms that is actually responsible for capturing and disarming any incoming free radicals. (Because of this, glutathione is abbreviated by chemists as GSH.) Once these damaging free radicals are removed, DNA, proteins, and enzymes can properly carry out such multitudinous functions as keeping the liver from getting toxic or destroying carcinogens in the blood. More research is needed to fully understand the healing powers of the cystine in AFA. Presently there is little incentive on the part of pharmaceutical companies to explore it because it cannot be patented as a drug.

Since the essential amino acid methionine is a precursor to cystine, the sulfur-containing methionine in AFA can also be used to help biosynthesize more. The mineral selenium is also important to the proper function of the antioxidant GSH. The selenium in AFA is easily assimilated because it is chelated in the form of the molecule selenomethionine or selenocysteine.

The extremely rich reserves of betacarotene (and glutamine) in AFA benefit our gastrointestinal tract, stimulating specialized cells to secrete beneficial lubricating substances (mucopolysaccharides). They help to form a protective barrier that guards us against the intrusion of invading bacteria and unwanted toxins, and they help us to absorb and assimilate amino acids more efficiently. After all, undigested protein is ultimately broken down by intestinal bacteria into dangerous and foul-smelling toxins that may eventually lead to cancer.

Selenium levels are typically low in rheumatoid arthritics. The small amount available in AFA algae is enough, on a daily basis, to activate GSH peroxidase. Its immediate effect, especially when used in conjunction with vitamin E (don't worry – it is already in AFA in the right proportions), is to be anti-inflammatory, thus relieving some of the arthritic symptoms.[4]

Numerous studies have confirmed the fact that as people (or animals) age, their sulfur-containing amino acids (methionine and cystine) begin to diminish.[7] Cysteine (a form of cystine that is much more absorbable) also efficiently deactivates free radicals and thus preserves and protects cells.

Osteoarthritis is a degenerative bone disease which probably effects 75 to 80 percent of people over age 50 and affects women ten times more than men. That adds up to a lot of suffering people. Unfortunately, Ibuprofen (Motrin) and other nonsteroid pharmaceutical drugs only cover up the symptoms while speeding up the disease. The natural methionine found in AFA may help to postpone the arrival of this common condition.

Free radical damage also greatly impairs the health of our gums. Our own antioxidant enzymes, such as GSH peroxidase, SOD, and catalase, are helpful. Unfortunately, excessive mercury accumulation greatly reduces their ability to function properly. AFA provides a powerful combination of antioxidants which synergistically work together to reduce the surprisingly high amount of free radical damage done to the gums.

Superoxide Dismutase (SOD) for Supple Tissues

This large SOD enzyme is also universally present and is probably the best antioxidant known. It is useful in keeping cell membranes and tissues fluid and supple. Because SOD is vitalized as an enzyme by the minerals zinc or copper, a shortage of any one of these can impair its full antioxidant abilities to prevent premature hardening of tissues.

> Although SOD works well when injected directly, most is destroyed in the gastrointestinal tract during digestion. But SOD is still a good source of amino acids and minerals from which human cells can reconstruct it. As its name implies, SOD is designed to disarm the superoxide ion, one of the most damaging of the oxygen free radicals. Tobacco smoking causes free radical damage to the cells mostly on the surface of the gums. This may lead to gum inflammation (gingivitis), which perhaps affects 50 percent of those over 50 years of age. The antioxidants in AFA algae help to replenish those that have been used up in quenching the free radicals generated by the smoke.

Manganese – The Ancient Bioactivator

Three billion years ago or so, blue-green algae such as AFA learned how to take the hydrogen out of water (leaving oxygen) and combine it with the CO_2 in the air to biosynthesize carbohydrates. This aspect of photosynthesis (discussed earlier in Chapter 2) is catalyzed by a manganese-containing enzyme that is able to temporarily bond to the water, separating its hydrogen and oxygen. However, when water is "split" in this manner, most of the oxygen gas needs to be quickly released into the atmosphere before it damages the cell membrane due to its dangerously high chemical reactivity. Indeed, whereas O_2 is used to oxidize carbohydrates for energy, excessive oxygen, especially as free radicals, can oxidize, peroxidize, burn, rust, and otherwise ruin delicate cellular biomolecules.

One way AFA reduces such destruction is through the superoxide free radical quenching abilities of manganese-bioactivated SOD. By ingesting AFA, we can utilize its manganese (and its amino acids) to help reconstruct SOD for the protection of our cellular membranes and nerve tissues.

Phytochemical Antioxidants in AFA algae

All plants and algae use one or another assortment of colored plant pigments such as carotenoids or bioflavanoids to protect them from free radicals. There are thousands of such phytochemicals, each of which is tissue specific and thus concentrated in one part of the organism or another. Irritable bowel syndrome, Crohn's disease, and other intestinal inflammation disorders can be partly irritated by free radical damage. Antioxidants in general, and AFA in particular, should help reduce free radical damage.

Betacarotene and chlorophyll are present in relatively large quantities within the photosynthetic bioantennae of blue-green algae. Although AFA contains a wide assortment of other carotenoids, often

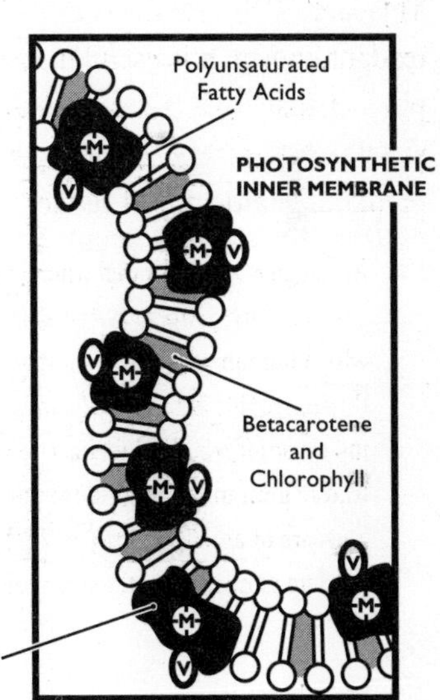

with more potency than betacarotene, the larger concentration of betacarotene, when ingested, provides profound protection for the health of bodily tissues such as the skin, eyes, and even the thymus gland.

Perhaps the most important property of betacarotene is that it seems to increase the life span of humans dramatically, especially if other carotenoids are present. Looking for more benefits, researchers at Tufts University reported that the antioxidant abilities of carotenoid compounds protect the eyes against free radical-induced cataracts[6]. Here is still another good reason for consuming AFA on a daily basis.

Folic acid is a nutrient growth factor for the lactic acid bacteria of the gastrointestinal tract. Vitamin B12 from AFA ingestion is also helpful. According to a 1996 study at the Center of Aging, Karen Riggs and her colleagues at Tufts University have shown that older men with deficiencies in folic acid and vitamin B12—which increase the risk of heart attack and stroke—are not able to perform well on specialized IQ tests and scored as poorly as people with mild Alzheimer's disease. Apparently when the heart is in danger, the mind does not work as well. The extraordinarily high amount of vitamin B12 in AFA thus helps the heart and mind.

Chlorophyll has been called a cell regenerator essentially because its central magnesium atom plays an important role in so many (325) different enzyme systems. The chlorophyll in AFA algae may be one of several factors in AFA that inhibits the mutation-promoting effects of environmental carcinogens. Chlorophyll may do this by acting as an antioxidant and thus protect our DNA during cell division, a very vulnerable time in the life all of our cells.

Chlorophyll is structurally similar to the hemoglobin in our blood and seems to help oxygenate the blood. It is also believed to increase peristaltic motion, soothe inflammation, reduce the severity of migraine headaches, reduce excess pepsin secretion and thus gastric ulcers, and even relieve chronic constipation. Heavy menstrual periods have already been successfully treated by chlorophyll supplements of only 25 mg per day,[7] an amount easily obtainable from a small amount of AFA.

Nitrogenase and Molybdenum – Their Enhancement Properties

AFA is capable of transforming molecular atmospheric nitrogen, N_2, into NH_3 (ammonia) by the use of an unusual enzyme called nitrogenase. This enzyme actually consists of two proteins linked together by chemical forces, the larger of which contains two molybdenum atoms and a variable amount of iron.

When we ingest AFA, its molybdenum can be used to activate several human enzymes associated with longevity-enhancement properties. Xanthine oxidase, for example, is an enzyme that produces small amounts of uric acid, which is capable of acting as a free radical scavenger.

Nitrase reductase is a molybdenum-bioactivated enzyme that converts nitrate ions into harmless amine functional groups. Without this enzyme, carcinogenic nitrosamine molecules would be produced, possibly leading to colon cancer.

There's even a molybdenum-bioactivated enzyme (sulfite oxidase) that oxidizes any toxic sulfite ions in our bodies, changing them to harmless sulfate ions. Sulfite ions are toxic because they destroy B vitamins and thus contribute to nervous system disorders.

Chromium Protects the Pancreas

The essential mineral chromium is known to enhance longevity factors, not because it is part of an antioxidant system, but because it plays a crucial role in the utilization of glucose and the metabolism of carbohydrates. Chromium is typically at a low level in the tissues of the elderly. When in high enough concentration, this mineral ion activates a substance of unknown structure called the "glucose tolerance factor," or GTF, which allows blood glucose molecules to more easily pass through the cell membrane into its cytoplasm.

Our American diet of super-refined foods robs us of this invaluable trace mineral. Ingesting AFA algae can help to alleviate some of the discomfort associated with diabetes and contribute toward decelerating the aging process associated with hardening of the arteries.

Because the chromium in the GTF of AFA helps us to assimilate glucose, we need not place dangerous demands on our pancreas to oversecrete insulin. And, just as too much oxygen is bad for our cells, too much glucose is also dangerous. High levels of this basic molecule can actually damage the shape, and thus the function, of proteins, peptides, enzymes, hemoglobin, and nucleic acids.

Healing Some Modern Diseases With Immuno-stimulating Minerals

ALZHEIMER'S DISEASE is characterized by the slow destruction of brain cells. This disability has already reached epidemic proportions. Approximately 15 percent of the elderly in the United States, about 3 million people, have mild to severe Alzheimer's disease. About 2 out of 3 nursing home patients have this disease. One consequence of old age appears to be a significant decrease in the production of protective antioxidants. These and other abnormalities in the immune function may begin to be reversed with zinc,[8] especially since the elderly seem to have an impaired ability to absorb this important mineral.[9] The zinc in AFA is highly absorbable because of the amino acids that are chelated to it.

OSTEOARTHRITIS, tied to the natural aging process, occurs when an increase in enzymes that tear down joint collagens outweighs the ability of the body to restore it. Vitamins A, E, and B_6 along with the minerals zinc and copper act together to synthesize joint collagen and cartilage structure. A deficiency in any of these may begin the process of accelerating the affects of osteoarthritis.[10]

OSTEOPOROSIS is another interrelated medical problem affecting mostly women. It is partly characterized by urinary losses of calcium and the resultant effects of bone breakdown. There are numerous dietary factors that put us at risk for this disease. Even excessive sugar consumption (150 grams per day) can worsen the effects and cause an increase in calcium excretion. Careful chemical analysis of the tissues between human cells (intracellular regions) in hypertense people with high blood pressure shows definite magnesium deficiency and relatively low values of calcium. Magnesium may even be more important than calcium in lowering blood pressure.[11] AFA is high in both minerals.

Essential Fatty Acids – Truly Essential to Membrane Fluidity

As we grow older, our individual cell membranes begin to loss their flexibility. As a result, the cell's most basic biochemical processes do not take place with the same fluid efficiency. Life-sustaining biochemical traffic, in which important molecules and ions squeeze back and forth across the cell membrane, begins to slow down as we age. As the roads of life begin to narrow, nourishment is not taken in as well and toxins are not expelled as quickly.

We have already seen how our antioxidants work around the clock to protect this magical membrane from damage. A second factor in maintaining youthful and vibrant cells, however, is to ensure that their cell membranes are constructed of the finest materials with the most flexible polyunsaturated essential fatty acids. If the fatty acids within the cell are clogged with cholesterol or if the fatty acids are saturated with too many hydrogen atoms, cellular rigidity begins and the aging process is accelerated. Dietary saturated fatty acids are often peroxidized to the point of being slightly rancid. When such fatty acids are incorporated into the cell membranes of otherwise healthy cells, the structure and function is impaired. AFA ingestion is one way of starting a regimen to improve the fluidity of our cell membranes.

When "bad" LDL cholesterol molecules in the bloodstream are oxidized by free radicals, white blood cells attack these molecules, engulf them, and then turn into a gooey mass of "foam cells" and die. This invariably clogs our blood vessels, dramatically raises blood pressure, and increases our risk of heart attack. The wide assortment of natural carotenoids in AFA (including alpha and betacarotene) can help prevent such heart-related problems by protecting the LDL cholesterol from free radical attack.

Providentially, AFA has no cholesterol in it, and a relatively high percentage of its fatty acids is unsaturated.[12] Essential fatty acids in AFA like linoleic and linolenic acids are present to help insure membrane fluidity. This should not surprise us. AFA algae needs to be as flexible and resilient as possible to survive the harsh conditions of Upper Klamath Lake, its indigenous habitat.

AFA possesses significant amounts of linolenic acids. A 1994 study reported in *The Lancet*,[13] which involved more than 600 patients, clearly demonstrated how these essential fatty acids helped prevent coronary heart attacks and premature death. In general, gamma-linolenic acid (GLA) is enzymatically biosynthesized from linoleic acid, a true essential fatty acid. The GLA can then be converted to a prostaglandin, which has a variety of propitious hormonal effects.

As artery wall plaque begins to form, sending blood pressure up, less oxygen is able to get to the heart. This is the beginning of atherosclerosis, and a first step toward angina pectoris. Long chain and free fatty acids (FFAs), which are the metabolic fuel for the heart function, begin to accumulate in the heart. Carnitine normally transports such FFAs into organelles within the heart cell called mitochondria (a heart within a heart cell), where they are burned for energy without the toxic side reaction of fatty acid metabolite production.[14] Fatty acid metabolites are known as "cell membrane disrupters" because they impair heart muscle contraction.

As the concentration of HDLs rises and that of LDLs decreases, cholesterol is being broken down faster than it is being deposited in tissues. Meanwhile, platelet aggregation is expected to decrease.[15] Put another way, the higher the HDL / LDL ratio the lower the effects of atherosclerosis.

The alpha-linolenic acid (ALA) in blue-green algae is fundamentally valuable in preventing coronaries. A 1994 paper in *The Lancet*[17] reports that patients who followed the experimental diet of consuming foods high in ALA had considerably fewer heart attacks than did the control group.

Caloric Restriction and Aging

It has been known for some time that laboratory animals live about 10 to 33 percent longer with fewer late-life diseases when placed on a low-calorie diet. In other words, the aging process is slowed down. Richard Weindruch and his co-workers at the University of Wisconsin-Madison are presently working to understand the biochemical reasons behind the phenomenon of the aging process.[18] The first thing to understand, of course, is that by ingesting fewer calories still means we must still get sufficient vitamins, minerals, proteins, and essential fatty acids. Many people who take AFA on a regular basis have reported that they seem to crave fewer calories and yet maintain a high level of ingested nutrients to maintain their bioefficient tissue operation and function.

Low-calorie diets (with an abundance of essential nutrients), according to Weindruch, can also postpone a variety of diseases associated with the aging process, such as cancer, immune system impairment, and problems of the gastrointestinal tract. As we age, our blood pressure and blood levels of both insulin and glucose rise while the sensitivity of our cells to take in glucose in response to insulin signals worsens. Low-calorie diets in monkeys show increased sensitivity to insulin and better glucose uptake. Many of the people of Okinawa consume low-calorie diets with abundant essential nutrients, much of which originates from algae sources, both direct (such as kelp and nori) and indirect (fish that feed on algae). The incidence of centenarians on Okinawa is, as noted by Weindruch's group, 40 times higher than on any other Japanese island. We'd love to see the ARP doing in-depth articles about these people.

To explain these important observations, we must look deep inside our cells, into the tiny cytoplasmic structures called mitochondria, those vital "power factories" that supply us with life-sustaining energy. Apparently, to live longer we need to minimize free radical injury to the membrane lining of the mitochondria. This is because more than any other part of the cell, the mitochondria are usually the hardest hit and most affected by such damage.

Salutary mitochondria are able to supply our energy needs because oxygen, enzymes, and other complex molecules attached to the membrane of the mitochondria ceaselessly manufacture, from food nutrients, molecules of adenosine triphosphate (ATP), a kind of molecular currency that pays the energy price required for almost all other life-sustaining cellular processes.

But, as no bioenergetic process is perfectly efficient, dangerous free radicals are also produced as unavoidable side effects. Unfortunately, the molecules most vulnerable to free radical attack are those that synthesize ATP within the mitochondria. This is because there is no protein to act as a protective shield for the membrane. As a result, ATP production is even less efficient, more free radicals are produced, and the aging process is accelerated as increased cellular damage takes its toll. When cellular damage increases, the tissues and organs that they compose become less able to carry out their biological roles.

Of course, our bodies automatically try to enzymatically repair such damage. Over time, however, the aging process gains momentum and the upper hand. Apparently, sick and damaged cells require more ATP to repair them, thus robbing us of the energy to stay healthy. Sick cells make us even sicker as we lose energy to repair them.

What can we do? Obviously we need powerful antioxidant molecules – be they pigments or enzymes – as well as the vitamins and minerals needed to invigorate them. Probably the best natural and wild source of all such antioxidant shielding materials is within AFA algae.

Nucleic Acids – New Longevity Factors

The large molecules of DNA and RNA carry our genetic information and are responsible for coding the synthesis of our proteins and enzymes. These delicate nucleic acids are protected as much as possible from free radical damage by a virtual army of antioxidant molecular shields in the form of enzymes, phyto-chemicals, peptides, and vitamins.

Most chemists believe that such large and complex molecules, when ingested cannot survive the difficult journey through our gastrointestinal tract without being torn down into much smaller and less useful fragments.

Recent experiments, however, cannot be ignored. Laboratory animals, when injected with nucleic acids, live up to twice as long as those that are not so treated Although more research needs to be done, the chemical synthesis of exotic poly-nucleic acids are now developing into new frontiers of life extension.

The cytoplasm of AFA algae, where we find its DNA and RNA, is a rich source of potentially life-extending nucleic acids. Many people also report an increase in mental clarity and even memory enhancement. DNA is often and most easily damaged during certain phases of cell division. This is when it is most vulnerable to dangerous mutation. SOD – superoxide dismutase – is an antioxidant enzyme that lowers such risks. Unfortunately, our supply of such protective enzymes diminishes as we age. The rich array of antioxidants available in AFA can help supplement some of these losses of SOD.

We all long for the longevity of Methuselah, the patience of Job, and the constitution of a Hunza. Yet we must celebrate mortality as a continuation of the miracle of birth. This body that takes us to school, to play, to work, to worship, to family and to love deserves nourishment of unparalleled virtue for this journey of heroic proportions.

And whether you view nutritional blue-green algae as your fuel for today, the fuel for the new millenium, or for your next frontier, it helps to possess the same endurance food as an Olympic Decathalon champion!

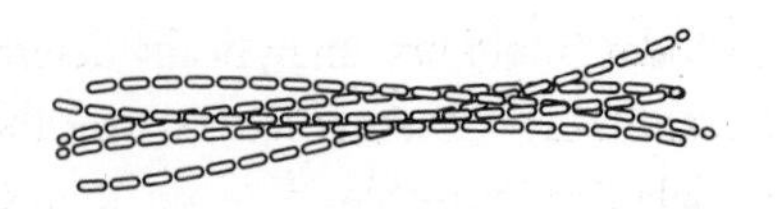

CHAPTER NINE

Mood Improvement and Amino Acid Therapy

Brain metabolism, neurochemistry, and the biochemical connections between amino acids and mood disorders are among the most exciting and revolutionary areas of nutritional research today. New and momentous discoveries point to how only twenty or so amino acids, individually or bonded together into neuropeptides, hormones, or proteins, create for us a kind of chemical language responsible for intercellular communication.

It appears, then, that our entire central nervous system is completely regulated by these chemical messengers. Of course, for one reason or another, there may be a disease or stress or dietary amino acid deficiency present that leads to a variety of neuropsychological imbalances. Usually a Pandora's box of the latest patented drugs are used to bring hope or, for some, relief, in spite of those familiar paragraphs of small print warning us of one too many side effects.

Now we are beginning to realize the wisdom of using natural medicine, that is, treating ourselves with the same molecular medicine with which our body treats itself. Many physicians today might first analytically determine if a deficiency in any important essential amino acid exists. If, for instance, there is found to be a low concentration of phenylalanine in the blood serum, then so many milligrams of that amino acid are prescribed to correct or reverse that condition. Hopefully, this is just the beginning of a more advanced and enlightened approach to dealing with mood disorders.

There are times, however, when a so-called megadose of one (or two) amino acids may bring on additional subtler side effects. For example, many amino acids compete for entrance into the brain, across the so-called blood-brain barrier. When a large amount of one amino acid is administered, the delicate balance of amino acid "brain traffic" becomes distorted. Some amino acids get metabolized too soon, others not at all. Chemical chaos may eventually begin to manifest as depression or dementia, a signal that brain chemistry is just not right.

One possible and logical solution might be to ingest amino acids in a proportion that the body understands – in relative amounts that the blood-brain barrier can smoothly assimilate. In 1984, the National Academy of Sciences[1] published an optimal amino acid profile showing the best recommended proportions of amino acids. Since our body cannot hold one amino acid until another comes along to bond to it, it becomes extremely important to insure, recurrently, that dietary amino acids be simultaneously *available and balanced*, or they will be wasted.

Two foods come closest to this optimum balance, the chicken egg and AFA algae. AFA has the most balanced profile of amino acids. For example, arginine and lysine compete with each other for transport from the intestine to the brain. When laboratory animals are fed high and unbalanced amounts of the lysine amino acid, there is a speedy reduction in the concentration of arginine in their brain cells.[2]

> The Food and Nutrition Board of the National Academy of Sciences has not, as yet, established RDA values for amino acids. Causes of depression are just beginning to be understood. There is mounting evidence that an imbalance in our amino acid intake may lead to an imbalance in monoamine function. A healthy balance of monoamines in the brain can also be disrupted by drugs such as nicotine, caffeine, and oral contraceptives. Monoamines include neurotransmitters such as serotonin, dopamine, and adrenaline.[3] The right balance of amino acids leads to an effective monoamine precursor therapy, which has been shown to be even more effective in treating depression than such conventional routes as the drug imipramine or electric shock treatment.[4]

Phenylalanine – At the Head of the Class

The essential amino acid phenylalanine is moderately present in AFA algae. It crosses the blood-brain barrier faster than any other amino acid. Because of this, it is highly concentrated in brain tissue. Phenylalanine is a precursor to tyrosine and enkephalin neuropeptides, both of which have mood-elevating qualities. Tyrosine itself is a precursor to the catecholamines, another important family of neuropeptides, which include dopamine and adrenaline.

Typically, phenylalanine is found in the blood as a single and free amino acid. It is converted to tyrosine in the liver. Most psychiatrists will openly admit that the drugs they prescribe to their depressed patients have many multifarious side effects. Tyrosine and phenylalanine are two amino acids found in AFA that show promise as alternative antidepressants.[5] Tyrosine levels are known to be low in depressed patients.[6] Phenylalanine is transformed in the body by enzymes to phenylethylamine, which has been long known to stimulate patients into improved moods.

The cells of our heart and brain do not live and die as other cells do. They are with us for the duration of our life's journey. Much of the micronutrients they use go toward the replacement parts they continually need in order to function. AFA provides the high-quality amino acids needed to build precious enzymes, and the essential fatty acids needed to maintain the flexibility of their delicate cell membranes.

Tryptophan – The Mood Improver

Tryptophan is the least abundant essential amino acid in food, yet it is probably the most widely studied amino acid in the world. Tryptophan is the sole amino acid that is transported by a blood protein, albumin, right to the blood-brain barrier, where it competes for brain uptake with branched-chain essential amino acids (BCAAs) and other amino acids.

What we now call depression or biochemical neural imbalance and what we once called melancholia is still a painful condition that affects many of us more than once in a lifetime. When tryptophan supplements are administered, the depressed patient seems to improve. Mood improvement and depression have been treated with amino acid tryptophan alone, with mixed results. Such studies have concluded that tryptophan therapy needs to be used along with other vitamin, mineral, and amino acid components. In fact, studies in Japan have had even more success by adding manganese ions (hopefully chelated) to the tryptophan, so as to accelerate production of the attention-worthy neurotransmitter serotonin.[7]

As the tryptophan level increases so does endorphin activity, with an increase in serotonin and a decrease in pain.[8] Some cases of overaggression in depressed individuals also seem to greatly diminish with tryptophan therapy, as well as a variety of other symptoms of schizophrenia. One small, but favorable side effect is a measurable increase in melatonin, a mood-elevating hormone secreted by the pineal gland, believed at one time to be the "seat of the soul."

Mood swings are closely tied to fluctuations in the biosynthesis of serotonin. Synthesis of serotonin in the central nervous system depends upon the amount of free tryptophan in the bloodstream. Typically, tryptophan transport in the brain appears greatly impaired in people suffering from depression.[8]

Lining the intestinal wall are millions of nutrient-absorbing finger-like protrusions (villi). During peak efficiency of the gastrointestinal tract —often brought on by ingesting AFA— highly specialized cells at the base of these "fingers" secrete a unique neurochemical messenger substance called serotonin that tells your brain that your intestines "feel good." In this way the brain knows when the body is being fed right.

AFA offers all needed components in a natural balance, making the tryptophan it contains far more effective. As a precursor to serotonin, tryptophan seems to noticeably curb carbohydrate cravings. This is because serotonin interacts with receptor sites on cells and actually increases the permeability of sugar across the cell membrane. Therefore, this increased sugar permeability provides the cell with needed carbohydrates.

During its biochemical journey toward creating serotonin, tryptophan also produces a metabolite, picolinic acid, which has been observed to relieve anxiety and even phobias. Another welcomed result of tryptophan absorption is that AFA's chelated zinc ions are better assimilated. It is interesting how often we see such a cascade of beneficial events, in which ingesting and balancing one family of nutrients leads to an enhancement of positive effects from another.

Pellagra is a common disease that results from a tryptophan deficiency brought on by ingesting too much corn. Diarrhea and dementia may result as low tryptophan levels typically lead to low levels of niacin. The tryptophan in AFA is readily metabolized into niacin, preventing common symptoms such as anxiety, insomnia, irritability, and depression, and concurrent complaints of memory loss.

Leucine – Natural Pain Reliever

Leucine, like isoleucine and valine, is a branched-chain essential amino acid that is probably active within small molecular neurotransmitters as well as being part of larger neuropeptides. There are also a variety of pain-relieving peptides called enkephalins, wherein leucine is often found. Enkephalins and the closely related endorphins often affect us by elevating our mood. Chronic fatigue syndrome is being studied with the use of short chain pentapeptides such as enkephalins and endorphins. When there is a bioimbalance of neurotransmitters, there is a tendency toward addiction and self-medication. This important observation is addressed in Chapter 11.

Isoleucine and leucine have consistently been measured at very low concentrations in the blood of extremely depressed children. Increasing one's intake of such BCAAs through natural sources like AFA may improve psychological moods over a period of time.

Methionine, Cysteine, and Substance P – Intercellular Communicators

These are the rare sulfur-containing amino acids that AFA does provide, but in relatively small amounts. They are often deficient in most diets without AFA, but certainly important for normal mental functions and positive moods. When mild or severe cases of psychosis begin to disappear, cysteine levels in patients steadily rise.

METHIONINE, an essential amino acid in AFA algae, is one of the two sulfur-containing molecules that is implicated in mood disorders. One interesting aspect of this valuable antioxidant is that it is involved in a wide variety of important biosynthetic reactions. Most of them involve the simple transfer of a carbon atom from methionine to dozens of other molecules that our cells require. However, if the body somehow senses a deficiency in this process, a number of depressive orders will typically be observed until the deficiency is corrected. Methionine supplementation has been shown to help depressed patients.

CYSTEINE, another intercellular communicator, is typically in short supply. Fortunately, it can be made in the body from methionine. A variety of emotional disorders are attributed to low cysteine/methionine levels. Apparently when these amino acids are in short supply, so is the biosynthesis of glutathione (GSH). Cysteine and the GSH that contains it are both important to the brain as neurotransmitters.

If levels of these amino acids are low, the body-mind connection communicates its need for us to ingest more amino acids. The amino acid content of AFA may well be one of the highest known. The good moods of people taking it on a daily basis may, in part, be traced to its unusually high percentage of AFA's digestible protein.

SUBSTANCE P: Mood enhancement and uplifting emotions sometimes can be related to a healthy and balanced attitude toward life. Good moods may also be connected to a balanced biochemical symphony of efficient neurotransmissions. A tiny protein or tetrapeptide called substance P plays an important role in moods and brain chemistry.

Substance P is a powerful neurotransmitter that is composed of arginine, lysine, and proline, all of which are available in AFA blue-green algae. The amazing effect of this neuropeptide is that it "sharpens" the mind as a learning promoter. It does this by stimulating brain cells to grow more dendritic spines or brain cell "arms" that can reach out to communicate with still more brain cells, thus enhancing our learning ability and in turn elevating our sense of self-worth.

Glutamine – DNA Synthesizer and Neurotransmitter

The nonessential amino acid glutamine is used in a variety of neurochemical reactions and, not surprisingly, is formed in the brain as a breakdown product (metabolite) of glutamic acid. Used in the synthesis of DNA, glutamine is also considered a major neurotransmitter and raw material fuel source for the brain. As expected, it is found highly concentrated in the brain, cerebrospinal fluid, and blood. Unfortunately, glutamine is not very abundant in most foods. However, in *aphanizomenon flos-aquae,* it is the *MOST* abundant amino acid, making it one of the best sources known for this "brain food."

Glutamine synthetase is a manganese-dependent enzyme used in the synthesis of glutamine. Since manganese is typically below recommended daily amounts in most people, glutamine concentration is also low. Here again, AFA is beneficial because its manganese content is useful in synthesizing even more glutamine.

The Lancet has reported[9] that glutamine is also useful in preserving the health and proper function of the gastrointestinal tract. A group of patients who did not receive any glutamine supplementation after surgeries in the hospital had a considerable decrease in the size and function of their intestinal villi, intestinal protrusions that absorb nutrients. Those who did receive glutamine had a healthier gastrointestinal tract, no decrease in villi size, and better intestinal absorption of nutrients.

Aspartic Acid and Asparagine – Energy for Neurotransmission

Aspartic acid and asparagine are both nonessential amino acids supplied in moderate amounts by AFA. Since they play a role in excitatory neurotransmission and provide the brain with energy, both are also rightfully associated with mood elevation.

One of tryptophan's metabolites is picolinic acid, which actually enhances the absorption of zinc and thus relieves inner tension, anxiety, and phobias.

Glycine and Serine – Calming Agents

Both of the nonessential amino acids glycine and serine also positively affect our mental state. Glycine, the simplest amino acid, may play a role in facilitating neurotransmitters like acetycholine and thus have a subtle, calming effect on the nervous system. Glycine receptors are found in cells throughout the central nervous system and brain tissues.

Serine supplementation has been known to diminish symptoms of psychosis. Does the body-mind connection seem to sense when serine is low and react accordingly? Perhaps. Serine, as phosphatidylserine, is a vital component of the phospholipid material of many cell membranes from a variety of different cell types.

Threonine–Reduces Neuron Dysfunction

Threonine, an essential amino acid, also seems to reduce some of the symptoms of depression. Threonine may do this in several ways. As an immunostimulant, the weight of the thymus gland has been observed to rise with increased levels of threonine. In addition to being a neurotransmitter precursor, threonine can also help to prevent neuron dysfunction.

Tyrosine – Important to Brain Nutrition

Tyrosine, a nonessential amino acid, helps to elevate moods, relieve depression, and increase mental alertness and memory. Tyrosine is able to do this because it is a precursor to dopamine, norepinephrine, and epinephrine neurotransmitters. Tyrosine is also a precursor to several hormones such as melanin, thyroid, and others. Its most important role, however, is to brain nutrition.

Vitamins, Essential Fatty Acids, and Mood Enhancement

Depressed patients often have a low concentration of tetrahydrobiopterin (BH4), a vitamin-like essential coenzyme used in the synthesis of several neurotransmitters. BH4 synthesis may be easily biostimulated with the help of AFA's plentiful vitamin B_{12}.[10] This biostimulation, along with a minor but synergistic contribution from folic acid and vitamin C, may partly explain why so many people who ingest AFA algae have reported such dramatic and almost immediate mood improvement. After all, many severely depressed and even manic patients have had complete remission of their symptoms by taking vitamin B_{12}, a micronutrient richly provided in AFA.[11]

Science News has reported that a study by Hilbben and Salem links a deficiency in decosahexaenoic acid to depression.[12] Seafood and AFA algae are primary sources of this polyunsaturated fatty acid because it can be biosynthesized readily from cardiovascular disease-reducing eicosapentaenoic acid, or EPA.

Let us not forget that lack of exercise[13] and depression are interrelated. In other words, AFA without cardiovascular stimulation – sports, walking, even sex! – just will not work quite as dramatically. Of course, appropriate psychiatric care, therapy, analysis, or spiritual practice must certainly be embraced.

CHAPTER TEN

Healthy Skin from the Inside Out

WHEN PEOPLE COMPLIMENT US BY TELLING US THAT WE look good and have a healthy glow, they are usually responding to how our skin appears. In other words, our skin can be and often is a revealing window to our inner nutritional health.

With a little background in nutritional chemistry they could just as well tell us how well our skin cells are being properly built, quickly repaired, and effectively protected. These are all important factors in skin health. If we carefully attend to these factors, our skin and overall health will improve every day, giving us that healthy glow we all deserve. AFA algae biostimulates collagen production and thus helps to give skin its youthful and elastic appearance. The proline, serine, and glutamine amino acids that are so well balanced in AFA are well-known for their collagen-stimulating properties. The vitamin A in AFA (and therefore the betacarotene that produces it) also helps to reduce the negative effects of skin thickening.[1]

Building and Repairing Healthy Skin Cells

We have learned that any healthy cell must have a fluid and flexible cell membrane. This, of course, allows for the smooth function of all cellular processes such as consuming nutrients or the speedy removal of toxins and wastes.

To build such a healthy cell membrane, we need the right raw materials. Fresh and organic fruits and vegetables, clean water, nuts, seeds, whole grains, and especially AFA provide our skin cells with the essential fatty acids, amino acids, and enzymes required for healthy skin cells.

Essential Fatty Acids

Essential fatty acids and other polyunsaturated fatty acids are needed to construct the bilipid layer of the cell membrane of our skin cells. One such essential fatty acid, gamma-linolenic acid, is bounteous in AFA. As we age, our ability to manufacture GLA from other fatty acids found in food begins to diminish.

Vegetable oils such as olive oil and uncooked flaxseed oil are rich in essential fatty acids, as are fish oils such as cod liver oil. Ironically, both oils may be taken orally to reduce excessive facial skin oiliness. Whales, salmon, herring, sardines, cod, and mackerel are rich in gamma-linolenic acid precisely because they dine on algae, their source of this life-sustaining cell membrane component. Eskimos who eat whale blubber seem to be immune to most heart diseases because the GLA it contains helps to maintain the health and flexibility of their cell membranes.

Amino acids in protein are needed to construct the proteins within and between the cells of our skin. Collagen protein holds our skin cells together, and elastin protein gives our skin its elastic and stretching properties. Proline and serine are among the important amino acids plentiful in these proteins, as are cystine and methionine, which contain sulfur. Enzymes are responsible for building and repairing cell membranes and the important proteins needed by all skin cells. As described in Chapter 14, there is a specific enzyme required for every step in the building and repairing of skin cells. To insure proper enzymatic function, specific vitamins and minerals are also required, all of which are generously available in AFA algae.

The lecithin present in AFA algae can help break down cholesterol bumps under the skin.

Essential fatty acids are necessary for the health of the skin. For example, the symptoms of eczema have been relieved by increasing one's intake of essential fatty acids supplied by AFA or evening primrose oil.[2] Typically, eczema sufferers have a deficiency of zinc-dependent enzymes and polyunsaturated essential fatty acids. These same fatty acids may also be supplied by fish oils derived from algae-eating fish. Whichever their source, the effectiveness of these essential fatty acids is explained by their inhibition of arachidonic acid, a nonessential fatty acid that leads to the production of inflammatory molecules called leukotrienes. Arachidonic acid comes from animal fats.

Red bumps and lesions around the neck and cheeks (seborrhoeic dermatitis) are usually caused by improper fatty acid digestion. Sometimes the faulty construction of our cell membranes may bring on this skin condition as well. In addition, if there is a biotin or vitamin deficiency in our intestinal bacterial, this condition will further deteriorate. Ingesting the eicosapentaenoic acid (EPA) found in fish oils and algae helps to inhibit the production of inflammatory leukotrienes, so skin problems such as psoriasis are greatly diminished.[3]

The proline, serine, and glutamine amino acids in AFA algae are well-known for their collagen stimulating properties.

Amino Acids

METHIONINE is uniquely important partly because it is what chemists call a "limiting" amino acid. This means that methionine is the least abundant amino acid in most foods. AFA actually has a high amount of this amino acid compared to many other foods. The antioxidant methionine in AFA is also a precursor to another nonessential sulfur-containing amino acid, cysteine.

CYSTEINE, when biosynthesized in the liver from methionine, also serves to detoxify the liver. Fewer toxins in the liver are ultimately beneficial to the skin.

In general, the sulfur in methionine and cysteine has always been nature's most important skin-enhancing mineral because it helps to keep our skin glowing with youthfulness and healthy clarity. Sulfur also helps to construct the protein keratin, which gives our skin firmness. Contained in both brewer's yeast and seaweed algae is a "skin respiratory factor," which increases collagen production in test-tube experiments. It is believed that this structurally unknown factor enhances the ability of fibroblast cells to take up oxygen and speed up skin repair in burn patients. More research is needed to decide if suspected sulfur-containing polysaccharides in AFA's cell wall are the same as or closely related to this mysterious skin factor.

AFA Vitamins for the Skin

BIOTIN is a B-vitamin coenzyme needed by a variety of enzymes (carboxylases) that, among other functions, build and repair skin cells. They do this by helping to carefully position and properly incorporate certain key amino acids directly inside various proteins. Our friendly intestinal lactobacillin bacteria synthesize it for us, provided we have a healthy bowel ecology. Chapter 13 discusses this important coenzyme further. Biotin seems to be useful in reversing eczema, dermatitis, and skin blemishes. Biotin is difficult to obtain and available only in trace amounts in brewer's yeast, whole grains, liver, and AFA.

FOLIC ACID is even more arduous to obtain. Because of the destructive effect of food processing on this relatively rare and unstable vitamin, folic acid may be the most common vitamin deficiency in the world. Since a variety of enzyme systems need folic acid for cell division, it is useful and essential in replenishing old skin cells with new ones. This keeps our skin looking healthy and feeling smooth with a minimum of wrinkles. Folic acid's name is derived from "foliage," hence good sources of this rare vitamin are uncooked leafy green vegetables. AFA also contains folic acid as do the friendly bacteria of our intestines. Folic acid is a complex and relatively large molecule that also contains PABA (para aminobenzoic acid), another skin-enhancing factor.

NIACIN (or vitamin B_3) also contributes to healthy skin. It is available in many meats and it can be synthesized from the amino acid tryptophan. AFA contains both niacin and tryptophan. These essential components work together in more than 50 enzymatic reactions to supply energy to skin cells, for skin support and repair purposes.

RIBOFLAVIN (or vitamin B_2) is a water-soluble B vitamin found in oily fish, nori seaweed, leafy greens, and AFA. It is also synthesized by our friendly intestinal bacteria. Vitamin B_2 is also involved in several enzyme systems that help our skin "breathe" by utilizing oxygen more efficiently.

PYRIDOXINE (or vitamin B_6) serves a variety of purposes because it is involved in about 60 different enzyme systems. Since many of these enzymes are used to build amino acids, which are then involved with skin-related proteins (e.g., elastin and collagen), B_6 can help enhance the health of the skin.

COBALAMIN (or vitamin B_{12}) has been called the "longevity" vitamin because it seems to increase energy, stamina, and activity levels, especially of the elderly. It is able to do this primarily because it helps activate enzymatic reactions that synthesize red blood cells. This reverses fatigue and seems to help motivate people to be more active and exercise more often. Thus, as we exercise and sweat more often, our skin rids itself of toxins and we start to look younger. Even acute and chronic skin conditions characterized by itchy welts over much of the body may be treated and possibly prevented with the vitamin B_{12} found in AFA.[4] A teaspoon (1.5 grams) of AFA algae provides about 200 percent of the RDA of this energy-giving vitamin. NO FOOD SOURCE IS HIGHER IN B_{12} THAN AFA.

PANTOTHENIC ACID (or vitamin B_5) is widely available in meats, vegetables, and algae (*pantos* is Greek for "everywhere"). It is thought to prevent skin wrinkles because it is involved with enzymes that produce cortisone, a stress reducer.

AFA Minerals for the Skin

IRON in AFA contributes to a lustrous skin tone because of the role it plays in the formation of the hemoglobin of our red blood cells. If our 20 billion or so red blood cells have their share of iron, each cell is better able to deliver oxygen to our skin cells and keep them healthy and radiantly fine-tuned.

CALCIUM activates an enzyme in the skin that repairs its cell membranes. This has a dilatory effect on skin aging. AFA is one of the most calcium-rich foods available. Thus, skin aging and calcium are interrelated.

CHROMIUM, found in brewer's yeast, has been used with some success in the treatment of acne.[5] This is probably due to the effect of its chromium content on glucose tolerance and insulin sensitivity. After all, acne can be viewed as a kind of diabetes of the skin. The chromium content of AFA along with its optimum amino acid profile may serve a similar purpose.

COPPER works synergistically with iron in several enzyme systems, which, among other functions, help remove cholesterol from our cell membranes. As a result, membrane fluidity increases and skin cells function more smoothly. Copper also assists several enzyme systems in the production of collagen and elastin, two skin cell proteins that contribute to the tightness and elastic stretching qualities of the skin.

MAGNESIUM deficiency may undesirably prevent the formation of enzymes that repair the collagen protein in skin. AFA algae is an excellent source of magnesium because of its high chlorophyll content.

ZINC is important in the treatment of acne because it helps to inhibit the production of enzymes that ultimately overstimulate the sebaceous glands of the skin, causing skin duct blockage and inflammation. The concentration of zinc in the blood of acne-prone 13-to-14-year-old boys is typically lower than any other age or gender group.[6]

Free Radical Protection for the Skin

Free radicals are unstable molecules with one (or more) unpaired and thus hyper-reactive electrons. Free radicals are a natural and very dangerous product of imperfect cell metabolism – natural, in that all enzymatic processes produce some free radicals as unavoidable by-products; dangerous, in that they continuously damage and alter biological tissue, bringing forth a variety of cellular dysfunctions such as cancer. Most free radicals are one of three unique forms of very destructive oxygen fragments called superoxide, peroxide, or hydroxyl free radicals.

Often, when one free radical strikes a cellular biomolecule, more free radicals are produced by chain reaction. This can become very dangerous. Because of this, all life forms from AFA algae on up have evolved an internal protection system of "antioxidant shields" to prevent such chain reactions from literally burning up and destroying every cell.

Without such antioxidant protection, there is an eventual deterioration of bodily tissues and organs as free radicals destroy cell membrane molecules (e.g., fatty acids, enzymes, proteins) and cell nucleus molecules (e.g., DNA, RNA, proteins). There is even some evidence that specialized genes within DNA called "longevity determinants" can be damaged and thus shorten our life span.

Superoxide Dismutase (SOD) – Antioxidant Enzyme

Superoxide dismutase, a ubiquitous antioxidant enzyme, protects us from the ravage of radicals in almost all cells. In humans, SOD prevents hardening of cell membranes, thus keeping our skin from aging too quickly. This interaction often enhances the youthful appearance of our skin until our enzyme effectiveness begins to diminish with age. Part of the reason for this is that SOD is bioactivated by any one of the three relatively rare mineral metals, copper, manganese, and zinc. The fact that all three are available in AFA is one reason it is so good for our skin. SOD has also been shown to protect our DNA during those especially vulnerable few minutes associated with cell division.

Around thirty years ago, superoxide dismutase was discovered in human cells by scientists John McCord and Paul Fridovich. SOD was found to enzymatically stop or "quench" many of the superoxide free radicals produced inside of us, making it a valuable "endogenous antioxidant." Later, Fridovich demonstrated how the most dangerous free radical of all, hydroxyl, could damage DNA and cause cancer. And soon after, Berkeley biochemists began to understand more clearly how free radicals can rip off their needed electrons from other important biomolecules, robbing each biomolecule of its function and usability.

Glutathione, Sulfur, and Selenium – Wrinkle Eraser

Glutathione is a tripeptide superoxide free-radical absorber that profoundly maintains the elastic youthfulness of our skin. It is also effective in protecting us from free radicals generated externally by the ultraviolet rays of the sun. Thus the incidence of skin cancer diminishes with ingestion of GSH. Since GSH is bioactivated by the presence of selenium, this mineral helps to protect not only AFA's cell membrane function but our own as well. It is probably the sulfur atom in the cystine portion of GSH that actually absorbs and disarms the incoming free radicals, making it beneficial to the skin.[7] Since methionine is a precursor to cystine, high methionine levels also contribute to keeping our skin healthy. Methionine is a sulfur-containing amino acid that is good for the skin as well as being a powerful free-radical deactivator.

The red blood cells of most acne patients are typically low in glutathione peroxidase. As expected, clinical trials have shown that the severity of acne can be somewhat reduced when treated with vitamin E and selenium.[8] This is because the proper function of the GSH enzyme is dependent upon the presence of both vitamin E and selenium. Their availability in AFA, along with a wealth of enzyme-building amino acids, may explain why AFA has anecdotally been reported to reduce acne severity.

Betacarotene as an Antioxidant Shield

Betacarotene, an orange-colored antioxidant molecular shield, is extremely powerful in protecting us from a variety of externally and internally derived free radicals. BGA contains more betacarotene than any other organism. Betacarotene protects AFA's cell membrane on the outside and its DNA on the inside so that future generations of blue-green algae are insured. By ingesting AFA's betacarotene, we share in this nutritional insurance, which protects us from cancer as well as keeping our skin beautiful. Betacarotene, with the help of chlorophyll, also protects some of the enzyme systems that are associated with the repair of skin

Dark lines under the eyes, called "allergic shiners," indicate a problem in nutrient absorption. Soothing components in AFA algae, such as chlorophyll, various carotenoids, and amino acid compounds, dramatically improve absorption right where it counts—where the outside world becomes the inside world—the walls of our intestines.

tissue. This factor naturally becomes more important with age, as does our need to ingest a very wide variety of different antioxidants with varying protective properties.

Betacarotene is called a precursor to vitamin A because it is used to biosynthesize that vitamin. Even Retin A, which many people have used as a "miracle" skin restorer, is actually a derivative of vitamin A. Both help to moisturize the skin and stimulate the growth of new skin cells. This antioxidant phytochemical also helps to protect the skin from environmentally-caused free radicals found in polluted air (e.g., smoke and smog) and water (e.g., pesticides and herbicides). Psoriasis may be controlled by ingesting betacarotene (and thus vitamin A) because it inhibits the formation of polyamine molecules, the generator of this skin disorder.[9]

Bioflavanoids Strengthen Skin Cells

Bioflavanoids are phytochemical polyphenols, which, among other roles, increase capillary strength. Strong capillaries allow nutrients to be more efficiently brought into each skin cell and help waste products to be more easily eliminated. This contributes to the healthy glow of oxygenated and detoxified skin. The bioflavanoids found in various berries (anthocyanidins) are similar to those found in AFA and are known for their collagen-stimulating properties.[10]

WITHIN AFA, bioflavanoids, antioxidants, minerals, vitamins, and amino acids all dance synergistically together. The very same substances that keep the cell membrane of AFA fluid and flexible are the same micronutrients which are needed for supple and healthy skin. The vitamins, minerals, and enzymes utilized by AFA during cell division are often the same substances that stimulate the growth of new skin cells.

It really should be no surprise to us that AFA is beneficial for the health of our skin. After all, the same micronutrients required by skin cells are the very same substances AFA uses to maintain, repair, and build its own skin-like membrane while the vitamins, minerals, and essential fatty acids within the cell membrane impart the same flexibility to the cell membranes of our skin.

Even the antioxidants which AFA uses for protection are used by our skin cells for their own protection. Thus, as AFA helps to protect as well as nourish our skin cells, it removes their toxins and purifies each glowing, healthy cell from the inside out. In the next chapter, we discuss how this purification can also miraculously help us break out of the downward spiral of cravings, self-medication, and addictions.

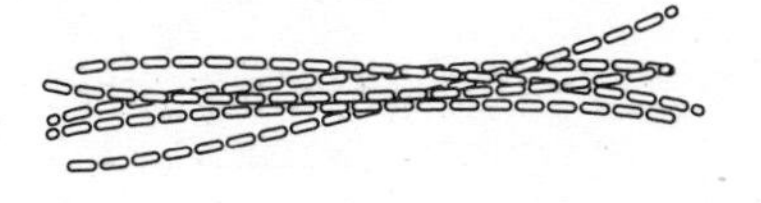

CHAPTER ELEVEN

Overcoming Addictions

SUBSTANCE ADDICTIONS ARE POWERFUL STATES OF PHYSICAL AND psychological dependence upon chemicals that ultimately rob a person of health and vitality as well as important moral intangibles. When we want to stop using or abusing drugs, alcohol, and/or nicotine, we often experience unpleasant effects called withdrawal symptoms. Usually such withdrawal symptoms are an unavoidable yet necessary first step to overcome during the process of conquering our addictive behaviors. At the same time, the inability to handle this array of extremely uncomfortable withdrawal symptoms is the very reason so many people end up returning to their chemical and psychological addiction. Today, alcoholism affects over 10 million people and probably results in 200,000 deaths per year.[1]

To overcome our addictions, we need to first develop an inner strength through a deeper understanding of why and how we first began our addiction. This is a personal experience that usually requires the help of friends, relatives, therapists, medical doctors, and other intervenors. Equally deserving of our full commitment, and parallel to ending our addiction, is the need to establish a powerful health-restoring nutritional regimen so that typical withdrawal symptoms, often quite horrifying, can no longer drag us back to addiction and loss of self-control.

AFA algae, along with probiotic and enzyme therapy (see Chapters 13 and 14), gives us the nutritional support to successfully cultivate a life without addiction and to establish a state of enduring health that will give us back the vitality we need to become a fully creative and inspiring, caring human being.

Essential Fatty Acids – Deficient in Alcoholics

Alcoholics are known to have a deficiency in essential fatty acids, especially those of the Omega 6 variety, because of prostaglandin excretion. Alcohol is broken down in the liver by specialized enzymes (dehydrogenases) into poisonous small molecules called acetaldehydes and other undesirable long-chain organic acids. Further complications typically involve fats deposited within the liver and mood swings related to increased hypoglycemic reactions.

Amino Acids and Maintaining Stable Blood Sugar Levels

One very general observation is that most people involved with an addictive disease are probably protein deficient. The easily absorbable protein available in AFA helps to give the addict an opportunity to maintain stable blood sugar levels. This then leads to a stable and more comfortable emotional state wherein natural neurotransmitters, such as serotonin and dopamine, replace the artificial chemical high from alcohol, drugs, and nicotine.

Nutritional bioactivation from a food source such as AFA will help to reverse the deleterious effects of chemical addiction. AFA will also help to reestablish a healthful homeostasis, which will protect the addict from the physiological depression and other characteristic symptoms that so typically return the addict back to self-medication.

Alcohol increases intestinal permeability to toxins[2] and undigested foods, causing allergic reactions that greatly increase and form immune complexes that deposit in body tissues. This chain of reactions increases cravings.

Alcoholism impairs the proper function of the liver, within which amino acid metabolism takes place. As a result, depression often sets in, partly because the amino acid content of the blood is so abnormal, far different than the balanced amino acid profile exemplified by a healthy human. The alcoholic is greatly benefited when normal amino acid levels are restored[3] by eating foods such as AFA algae with its balanced amino acid profiles. Restoration to normal levels helps the patient, while branched-chain amino acids inhibit brain dysfunction.[4] BCAAs in AFA help to inhibit the breakdown of proteins that accompany cirrhosis of the liver brought on by alcoholism.[5] Relatively small amounts (1 gram per day) of glutamine (a nonessential amino acid) have been used in human and animal studies to reduce alcohol cravings.[6]

Amino acids have been used to fight drug addictions. Methionine for heroin, tyrosine for cocaine, and glutamine for alcohol have been used to combat addiction. Detoxification programs have utilized amino acids such as cysteine and glycine. Along with B vitamins, antioxidants, chlorophyll, and essential fatty acids, a variety of detoxification and withdrawal symptom-lowering regimens have been developed. If we look carefully at all such detox programs, we begin to see how the ingredients within AFA are being used in a variety of natural approaches in drug treatment programs.

Glutathione as a Detoxifier

Glutathione is synthesized partly from cysteine, a sulfur-containing amino acid found in liver, wheat germ, garlic, and AFA. Antioxidant tripeptide glutathione (GSH) may help to reverse cirrhosis of the liver brought on by alcoholism. In the form of GSH, heavy metals such as lead and cadmium can be chelated and eliminated. Cadmium toxicity often occurs in the lungs of cigarette smokers and is one factor known to significantly reduce sperm count in men.

GSH helps to decrease the toxic effects of unnatural chemicals and drugs in general because of the role it plays in a variety of enzymes that break down such chemicals in the liver. GSH is also a critical and important antioxidant that protects those parts of the body most vulnerable to the free radical toxins and pollutants of alcohol, drugs, and nicotine. Cysteine seems to be particularly protective against cigarette smoke and alcohol. The immune system of heavy smokers is characterized by low numbers of thymus helper cells. Fortunately, these T-cells begin to return when smoking is stopped.[7]

Tryptophan – Nature's Tranquilizer

Tryptophan, an essential amino acid found in AFA, has been used as a natural tranquilizer because of its role in the biosynthesis of serotonin, an important mood-elevating neurotransmitter. One reason alcoholics have a high suicide rate[8] is believed to be that serotonin metabolism is related to depression.[9]

As a serotonin precursor, tryptophan may have some effect in reducing alcohol cravings, and of some other stimulant drugs. Since serotonin production in the brain is a psychological reward of sorts, it is possible that drug cravings and the need to self-medicate would correspondingly diminish.

Nicotine is a drug known to stimulate the adrenal glands to secrete hormones that greatly inhibit the uptake of serum tryptophan, thus resulting in decreased serotonin brain activity,[10] which can lead to depression. Cigarette smokers who eat AFA algae and then quit smoking will experience fewer withdrawal symptoms due, in part, to the tryptophan in AFA.

Alcoholics have very little tryptophan left in their bodies, and what little they do have is barely able to cross the blood-brain barrier to form the neurotransmitter serotonin.[11] This explains, in part, the sleep disturbances and depression symptoms experienced by alcoholics. In fact, low levels of tryptophan may lead to coma or brain damage, commonly known as "wet brain."[12] The tryptophan in AFA, along with its 19 other amino acids, helps to elevate the mood and improve the sleep patterns of the recovering alcoholic.

The dietary requirements for niacin may be met by the ingestion of tryptophan, as found in AFA. When niacin is up, serotonin is up, consequently anxiety, depression, and carbohydrate cravings are diminished.

Tyrosine Reduces Drug Withdrawal

Tyrosine, a nonessential amino acid found in AFA, is also involved in the manufacturing of neurotransmitters that are derived from phenylalanine. Tyrosine has been used in the treatment of drug withdrawal. When used with tryptophan, the combined mood-elevating effects seemingly reduce depression and anxiety, making drug withdrawal less daunting.

Leucine, Isoleucine, Valine – Help for the Liver

Leucine, isoleucine, and valine are the branched-chain amino acids that many body-builders use to help muscle production. They have also been shown to be helpful in treating liver damage caused from excessive alcoholism.

AFA Minerals for Enzyme Bioactivation

Calcium levels tend to drop as our consumption of nicotine, alcohol, and coffee increases.[13] As a result, smokers will generally have a much lower mineral content in their bones. A physical exercise regimen will help to prevent the enormous increase in calcium excretion that accompanies the dangerous combination of drugs coupled with low physical activity.

Magnesium – Vitality for a New Life

Magnesium is a mineral ion that is absolutely essential to the life of any cell. In fact, the magnesium ion content of AFA is about equal to that of sodium. Magnesium aids in vitalizing and biostimulating enzyme systems, helps to deepen the elec trical currents balanced inside and outside the cell, and, with calcium, regulates th rhythm of our very heart beat.

It is no wonder, then, that a magnesium deficiency often leads to subtle bu pervasive depression, low energy, an inability to concentrate, decreased appetite and even suicidal thoughts. These kinds of deficiencies – which can be correcte with leafy green vegetables and AFA algae – are common, especially in the elderl population. Symptoms such as these often lead to addictive alternatives that star with self-medication and all too often lead to alcoholism and drug abuse in ar attempt to avoid difficult bouts of depression.

Enzymes Bioactivated With Selenium

Alcoholics typically have deficiencies in minerals such as selenium that activate antioxidant enzymes such as GSH peroxidase. Because of this, there is increased peroxidation of the phospholipids of the cell membranes of alcoholics.[14] Selenium is an important trace mineral that, even in extremely small quantities, is essential to all humans for survival. AFA uses it as we do, as part of the antioxidant tripeptide glutathione (GSH). Selenium in GSH protects us from cancer, detoxifies heavy metals, and even biostimulates the immune system. As a detoxifier, selenium is useful in rendering harmless the otherwise dangerous effects of rancid or peroxidized fats, which are caused by a variety of free radicals. This boosts the immune system because it ultimately enhances the health of the cell membrane.

As alcohol consumption goes up, liver cells begin to die off because of the damaging effects of free radical peroxides on the flexibility of these cell membranes.[15] Without some kind of antioxidant protection (generously provided by AFA), fatty infiltration of the liver begins to take its toll.[16]

Zinc – The Most Important Enzyme Bioactivator

The long road to addiction recovery must include a thorough detoxification program. The most important enzyme systems for alcohol detoxification are mostly dependent upon zinc.[17] It is an established fact that as the amount and duration of alcohol consumption goes up, the concentration of zinc goes down.[19] This is due to the inability to maintain proper intestinal zinc absorption with chronic alcohol consumption. Once this happens, enzyme synthesis function decreases, followed directly by the inability to synthesize proteins to maintain cellular repair. Because of its zinc, amino acid, and essential fatty acid content, AFA is highly recommended as an integral part of any detoxification program.

Vitamin Needs for Addicts and Alcoholics

Vitamin deficiencies in addicts are one of the major causes for their ill health. The presence of nocuous intestinal microflora in alcoholics[20] lead to a toxic bowel, which greatly impairs food absorption and vitamin assimilation, especially of vitamin B_{12} and folic acid. Betacarotene, vitamin A, zinc, and folic acid are common deficiencies, which often lead to a profound decrease in immune and liver functions, as well as a variety of other complications.

Almost all alcoholics are deficient in vitamin C, one of the more important substances needed to help eliminate excess poisons (acetaldehyde) generated in the liver.[20] There is much more vitamin C in AFA algae than in either *Spirulina* or *Chlorella*.[21] Alcoholics are typically unable to utilize most B vitamins, either in the food they eat or in the supplements they buy, especially if their drinking continues unabated.

Alcohol, heroin, and many other drugs will often, as expected, chemically reduce the absorption of most B vitamins in the intestines. Sometimes you can tell if you have a B vitamin deficiency by the appearance of a slightly swollen red tongue. Even tea and coffee drinkers can have this problem, especially the elderly. Correcting this problem by abstaining from alcohol, coffee, and tea can be almost immediately beneficial.

For example, when folic acid is properly absorbed, it can be used to begin the important process of detoxifying cells.[22] Folic acid also seems to help lower substance cravings, migraine headaches, and even mental disorders such as schizophrenia. Alcoholics and elderly people are at risk of folic acid deficiency because of general and intestinal malabsorbtion problems. Because of this, a combination of AFA, plant enzymes, and probiotics is recommended.

When alcohol breaks down, it turns to a smaller and much more toxic molecule (acetaldehyde) in the blood. One of the B_5 bioactivated enzymes speeds up the detoxification of this and similar toxins.

Choline Helps to Deal With Withdrawal Symptoms

The B vitamin choline is one of the components of AFA's cell membrane and is essential toward maintaining healthy cell membrane fluidity. Choline is also needed for the biosynthesis of important neurotransmitters, such as acetylcholine, which play a crucial role in mood elevation.

Because of this, we should not be surprised to learn that choline deficiency is intimately connected with memory loss and its ingestion may possibly help in preventing Alzheimer's disease. The choline of AFA is found mostly within its very flexible and accessible cell membrane.

Choline does seem to affect mood disorders, and studies have shown reasonably good improvement in relieving depression and stabilizing manic-depression. As with magnesium, getting enough choline from AFA and leafy greens helps people to better deal with addiction withdrawal symptoms. The choline available in some other sources, however, such as in highly saturated fats and oils, is not nearly as beneficial. Saturated fats tend to rob us of any membrane fluidizing effects, thus diminishing the effectiveness of the choline. Studies under way now are looking at morphine addiction withdrawal and recovery. As expected, specialized drugs that contain choline and polyunsaturated fatty acids are proving to be highly promising.

Thiamin – Important in Recovery

Thiamin (vitamin B_1) is unfortunately among the most typically deficient vitamins in addicts. Alcoholics are especially at risk for the harmful effects of having their vitamin B_1 biochemically deactivated. Since this vitamin (along with B_{12}) is essential in changing blood glucose into usable energy for producing red blood cells and maintaining nervous tissue, vitamin B_1 deficiencies often produce mental and emotional confusion. Sometimes there is even difficulty in walking. Because alcohol blocks proper absorption of this important vitamin at the walls of the intestine, it is very important that recovering alcoholics get their B_1 vitamin from a wholesome, natural source to feel normal again as soon as possible. Since the thiamin in AFA algae is chelated and naturally bonded to amino acids, its absorption rate into the human system is greatly enhanced.

Niacin – Old Toxin Remover

Niacin (vitamin B_3) does a wonderful job in lowering cholesterol, detoxifying pollutants and drugs, and stabilizing moods. Recovering addicts of all persuasions may benefit from its abilities to remove toxins that have stubbornly been stuck in fatty tissues. The niacin in AFA is chelated and thus easily assimilated. Cigarette smokers have relatively high blood levels of lead and cadmium (and low levels of vitamin C), as compared to nonsmokers. Deficiencies in such heavy metals, especially lead and cadmium, lead to hypertension and high blood pressure.[23]

Anyone who thinks they have a problem with drug or alcohol addiction is encouraged to seek counseling, get seriously involved in addiction recovery therapy, and seek and participate in an appropriate twelve-step program.

CHAPTER TWELVE

Easy & Safe Weight Management

MANY AMERICANS AS WELL AS EUROPEANS WANT TO LOSE weight, knowing they will feel better, look better, and live longer. Obesity and the conditions related to it, including diabetes, are among the leading causes of early death. Those of us who really want to diet and lose weight need to make some basic changes in our lives. Otherwise we might regain the weight we will have lost, and need to start over again. This lose-gain-lose cycle is a dangerous stress on our health and should be replaced with a safer, more gradual and long-lasting approach.

There are numerous systems and scientific approaches to weight management from which to choose. There are good points being made by all of them and their success rates vary, depending on many factors. One landmark study that should be kept in mind is a medical study at Tulane University in 1971, which showed that when animals were overfed while young, their fat cells multiplied so quickly that the number of fat cells retained as an adult was permanently greater, sometimes by 300 percent. Obesity results when there are more fat cells to feed. Dieting will be more difficult and should be more gentle for those who have acquired more fat cells from youth.[1]

Dr. Atkins and Dr. Stillman have their special diets, and there also is the much-criticized Scarsdale Diet™. Weight Watchers™, Jenny Craig™, and others offer their franchised points of view as well. If we look very closely at their philosophies and approaches, however, a pattern emerges that begins to look more obvious, remains challenging, and heavily relies on sensible nutritional supplements, all of which, we believe, are available through ingesting AFA and supplemental enzymes.

Processed, canned, and overcooked foods do not contain any inner (endogenous) enzymes of their own because the proteins that form their "enzyme backbone" have been hopelessly altered and denatured by the heating process (see Chapter 14 for details). Canned – and therefore enzymeless – foods often overstimulate the entire endocrine system, promoting weight gain.[2] In general, cooked food calories actually cause weight gain more readily than raw food calories. As one study observed, "Technical men in the business of extracting the maximum profit from farm animals found it was not economical to feed hogs raw potatoes. The hogs would not get fat enough. Cooking the potatoes, however, produced the fat hogs that brought the farmer the kind of money required to make a fat profit."[3]

Some physiologists believe that when the hypothalamus of the brain is constantly being stimulated by refined carbohydrates, obesity may result. This may be interpreted from a classic British study that showed how laboratory rats steadily increased their percent of fat when fed their typical diets of processed and enzyme-free proteins, fats, and carbohydrates.[4]

Tests on obese patients at Tufts University School of Medicine showed how overweight people may be deficient in lipase, a fat-digesting enzyme that also helps in the overall metabolism of dietary fats.[5] Eating fewer processed foods, coupled with a diet containing AFA algae and enzyme supplements, may begin to reverse this trend toward obesity. Processed milk products, especially butter, should be avoided because they lack lipase and may therefore allow cholesterol deposits to accumulate on artery walls. AFA, with its abundance of enzyme-activating amino acids, minerals, and vitamins, in concert with enzyme supplements containing lipase, may help to reverse this life-threatening trend as well.

The Optimum Way

All weight management systems have the following general aspects in common: New habits or behaviors should be adopted that allow some foods to be eaten, and others, for one reason or another, to be avoided. New habits are slowly adopted. As their benefits become felt and reinforced, they become easier to maintain. Try the following:

1. *Start the day with probiotics.*
2. *Take digestive enzymes with 16 ounces of good water five minutes before eating.*

3. *Eat and chew food slowly, without too many technological distractions, preferably in the dining room, and with gratitude and calm.*
4. *Eat unprocessed foods and organically grown fruits, vegetables, nuts, and seeds, in salads and soups in early dinners. Reserve high-protein food for breakfast and lunch. Have an AFA unsweetened drink for breakfast.*
5. *Avoid meat (especially during dinner), fats (use only nonfat milk), salad dressing, desserts, processed foods, and creamed soups and sauces. Eliminate alcohol, caffeine, sugar, and salt. See a nutritionist and a doctor for details. Generally, eat less food.*
6. *Keep a good self-image by exercising physically, emotionally, mentally, and spiritually. Don't shop for food when hungry; indulge and break the above rules* ONCE *per week.*
7. *Eat AFA in the amounts and during those times you find most effective for your own weight management. You may find that you eat less when you are nourished.*

The Easy Way

Ingesting the AFA

We strongly recommend eating AFA algae because its wide variety of nutrients helps us in two important ways:

- *AFA provides important vitamins, minerals, and other micronutrients to maintain our weight-loss program and to assist us in digesting, assimilating, and metabolizing our food efficiently.*
- *AFA reduces cravings for sweets and allergenic foods and generally suppresses the appetite, increases our energy, and improves our overall mood.*

HOW AFA'S MICRONUTRIENTS HELP A WEIGHT-MANAGEMENT PROGRAM.

Chlorophyll – The Cell Regenerator

AFA contains 2 to 3 percent chlorophyll, placing our blue-green algae among the highest on Earth for content of this wondrous green molecule. Chlorophyll is able to invigorate the intestinal lining and thus enhance the digestive and assimilative processes so that we are more nourished on the cellular level and thus have fewer food cravings. Chlorophyll increases peristaltic motion, it soothes inflammation, reduces excess pepsin secretion and gastric ulcers, and even relieves constipation.

AFA has been called a cell regenerator partly because this green pigment molecule may also help to bring oxygen to the blood. Some biochemistry studies in Japan have shown that chlorophyll can do this because it is so closely related structurally to the hemoglobin of human blood that it may be able to chemically hold iron as does hemoglobin and thus use it to transport oxygen to each of our cells.

Fiber – Helps to Eliminate Toxins

AFA contains 5 percent fiber, ranking it high among all other sources of this valuable substance. Fiber greatly helps our colon to eliminate the extra toxic wastes excreted by our cells during any weight-loss program. Because fiber coats the intestinal tract, fat absorption is diminished. This is one reason why dieters ought to take about a gram of AFA algae (with 16 ounces of water) 30 minutes before eating. Of course, the water itself will help to suppress the appetite and along with the AFA will help the body METABOLIZE STORED FAT. We also recommend squeezing about a quarter of a lemon into the water to rouse the digestive system.

People with irritable bowel syndrome are typically deficient in vitamin K. The high chlorophyll content of AFA can be helpful.

Carnitine – Transports Fat

Although the amino acid carnitine is not found in AFA, it is found in very small amounts in some of the higher plants and in animal tissues. Since it is largely concentrated in animal muscle tissue, vegetarians often risk carnitine deficiency.

This amino acid is nonessential, however, because it may be biosynthesized in humans from lysine if there are sufficient amounts of the five or so enzymes necessary for its biochemical transformation: iron and manganese as mineral cofactors, with methionine (sulfur-containing amino acid) and vitamins B and C. If any of these enzyme-activating mineral substances are missing, the lysine to carnitine synthesis begins to slow down or even come to a dangerously unhealthy halt. AFA contains all of the necessary enzyme and coenzyme ingredients to help the body biosynthesize carnitine.

So long as one feeds

on food from unhealthy soil,

the spirit will lack the

stamina to free itself from

the prison of the body.

RUDOLF STEINER

Carnitine production is useful in weight management because of the role it plays in transporting fat across the cell membrane into the energy factories (mitochondria) of our muscle cells. Once there, that fat is oxidized and the energy is stored in an energy molecule (ATP) for quick use by our muscles. Because of this, we will see improved athletic performance and endurance as our fat is more efficiently burned for our energy needs. Also, we can exercise longer without fatigue and burn up more calories. In effect, carnitine production results in changing more fat into lean muscle. Carnitine also seems to detoxify the liver by stimulating the flow of bile.

Phenylalanine – Good Mood For Dieters

Another benefit of taking a gram of AFA thirty minutes before meals is that its phenylalanine has been shown to elevate the mood and then, in turn, suppress the appetite. This essential amino acid is able to interfere with the activity of enkephalinases, enzymes that break down enkephalins and other morphine-like endorphin hormones. Since these are mood-elevating neurochemicals, phenylalanine is connected with both pain reduction and a psychologically pleasant state that seems to reduce the need to seek out and crave foods in an attempt to elevate the mood.

Branched-Chain Amino Acids – Stimulates Muscle Production

Valine, leucine, and isoleucine are essential amino acids of similar molecular shape,

There are a wide variety of dramatically useful polypeptides available in AFA, which are directly absorbed into our bloodstream and dramatically enhance the permeability of the cell membrane. This is sometimes followed by an indisputable sense of well-being as nutrient flow within certain cells begins to improve. The fifty or so neurotransmitters are a kind of chemical language between and among the brain cells and other organs of the body. Their use is revolutionizing the treatment of psychiatric diseases and eating disorders.

which stimulate protein synthesis in muscle tissue. (They are called "branched" because the atoms of carbon in their molecules go off in three directions.)

These BCAA are used mostly by our muscles, especially during stress, and are often given to undernourished patients after surgery to help build muscle tissue. Leucine in particular and BCAA in general actually stimulates protein synthesis and inhibits protein breakdown. *Thus fat is metabolized faster than muscle.* This is especially desirable to weight lifters, muscle builders, and other athletes, and ought to completely replace the use of the inferior and dangerous steroids that have a laundry list of dangerous side effects.

Too much leucine, however, may sometimes lower and inhibit serotonin and dopamine production. This could aggravate psychosis in schizophrenics, but, more commonly, bring on "bad moods." During such moods, a self-medicating dieter often reaches for an overtly sweetened soft drink or pastry in order to experience temporary comfort from binging and eating too much high caloric food.

People who binge experience a relief of negative mood states. One of the psychological responses to massive food intake is a numbed state, which is similar to that achieved through drugs and alcohol. We cannot attribute compulsive eating or binging entirely to depression. Binging is also associated with trauma, sexism, low self-esteem, and many other factors.

Tyrosine Decreases Food Cravings

Tyrosine, a nonessential amino acid, may be a good way to naturally control and diminish the appetite. Found in moderate amounts in AFA, tyrosine works well with zinc to decrease food cravings. Compulsive eaters often binge in response to an emotional hunger rather than a physical need.

Tyrosine, a precursor to thyroxin, is an important thyroid hormone associated with reducing symptoms of depression. For this reason tyrosine helps to keep dieters satisfied and hence help them eschew the desire for more food to compensate because of a bad mood.

Tyrosene is also involved in the biosynthesis of neurotransmitters derived from phenylalanine. Tyrosine can be used in the treatment of drug withdrawal. Together with tryptophan, the combined mood-elevating effects seemingly reduce depression and anxiety, making drug withdrawal easier, and eating disorders easier to treat.

Arginine, the DNA Synthesizer

Arginine is an amino acid involved in DNA synthesis, lowering of cholesterol, and production of human growth hormone. It is present in relatively high amounts in AFA algae. It helps to curb the appetite in adults.

THE SYNERGISTIC COMPONENTS found in AFA biostimulate the body to establish its new balance – or homeostasis. This balance naturally diminishes our appetites and cravings while at the same time boosts our energy, so that our bodies might find their best and healthiest weights.

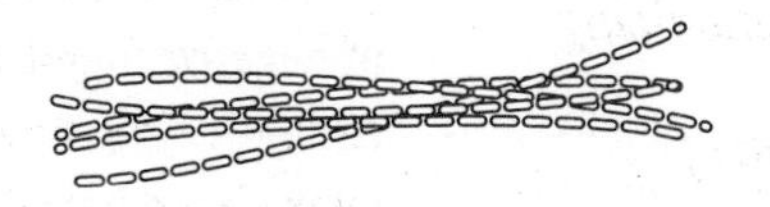

CHAPTER THIRTEEN

Probiotics – For Life!

Most of us are unaware of how our body stays healthy or how it tries to regain its health during times of sickness and disease. That is understandable; not everyone has the time or the inclination to study medicine, nutrition, or biochemistry. Yet there are certain simple and useful scientific facts that everyone should know.

One such fact is that your natural and wildly healthy gastrointestinal tract has always been the warm and cozy home of about 400 species of trillions of friendly, health-providing bacteria. Any disturbance of their normal bioactivity can have immediate consequences that affect our entire biological status. The job of these types of bacteria is to protect you from dangerous and invading bacteria and provide you with a variety of vitamins and other nutritional factors such as digestive enzymes. These bacterial allies considerably bolster the body's ability to detoxify a variety of dangerous carcinogens and build genuine intestinal fortitude. In return, these friendly symbionts are provided with food and a comfortable place to dwell. If you are healthy, they are happy.

Approximately one out of seven people in the United States has some form of intestinal dysfunction, most often colitus or irritable bowel syndrome. While the cause is unknown, pain, anxiety, and even a distended stomach often occurs. There is an incredible list of other symptoms, both chronic and acute, which are given later in this chapter.

The Dangers of Toxic Bowel

Our gastrointestinal tract is healthy only when it is being generously populated with a host of friendly microbial flora. Unfortunately, ever since the 1920s, when medical science hailed the advent of antibiotics such as penicillin and ampicillin as a medical miracle and a "magic bullet" that could cure almost all of humanity's ills, friendly and unfriendly bacteria alike have been destroyed in our gastrointestinal (GI) tract. As necessary as it may have been, we were not properly informed about the side effects of antibiotic therapy. We were not told how to replenish and repopulate our GI tract with new friendly species of health-sustaining microflora. Antibiotics is not the only way modern forms of medical overkill have destroyed our healthy probiotic gut. Alcohol, coffee, stress, and drugs of all kinds in our modern society have the same effect as do steroids, cortisone, too much sugar and alcohol, prednisone, and immune-suppressive drugs. Irritable bowel syndrome or Crohn's disease can allow specific toxic bacteria to destroy the cells that line the walls of the intestine, sometimes resulting in tragic consequences.

Proper elimination reduces the toxins and antigens, which may cause diabetes, ulcerative colitus, or thyroid conditions from the bowel bacteria. Fiber material present in AFA tend to bind to bacterial toxins in the bowel and then get excreted along with feces.[1] Along with a probiotic regimen, AFA algae's complex carbohydrates, balanced amino acid profile, and great diversity of minerals and vitamins (especially betacarotene) are therapeutically significant to reverse the symptoms of toxic bowel. Even premenstrual breast tenderness is dependent on bowel movement frequency[2] as well as *acidophilus* population.

Endotoxins are the cell wall components of certain unfriendly bacteria that proliferate in a toxic and dysfunctional bowel. These same bacterial toxins can cause an increased immune response (IgE and IgA immune complexes).[3] Such toxins can even depress the formation of cartilage tissue and thus increase the incidence of osteoarthritis.[4]

Migraine headaches may be caused by a dysfunctional bowel when the amino acid tyrosine is converted into tyramine by toxic colon bacteria.[5] Even sore throats and dizziness may occur. Unfriendly microflora can have the insidious effect of changing some compounds found in a constipated colon into toxic side products that could lead to colon cancer. Urinary tract infections (caused by proteus bacteria) can easily develop, as can the yeast overgrowth of *Candida albicans*, which many physicians connect with chronic fatigue syndrome, cystitis, prostatitis, vaginitis, athlete's foot, bloating, cravings, depression, and frequent colds. Even osteoporosis ("brittle bone" disease) is thought to be accelerated if our friendly microbials are not being treated kindly. What can we do?

The Joy of A Healthy Gut

This is where probiotic therapy comes in. The very word "probiotic" means "for life" in Greek. We can begin to reestablish healthy, living, friendly bacteria in our intestines by simply taking a few capsules of an assortment of friendly bacteria. Homemade "alive" yogurt can help, too. Depending upon our lifestyle, though, and level of chemical and psychological stresses, we will probably need to continue taking probiotic supplements. This will guarantee a gastrointestinal (GI) tract strongly populated with health-providing microflora. *Lactobacillus* bacteria produce a variety of B vitamins (especially B_{12}), biotin, folic acid, and vitamin K.

A healthy and beneficial GI ecosystem begins at birth where both permanent friendly indigenous bacteria (e.g., *L. acidophilus* and *Bifidobacterium*) and transient friendly bacteria (e.g., *L. bulgaricus*) populate the mouth and small intestine, with an especially high population (about 100 times more) in the large intestine.

Biotin is synthesized from our beneficial bacteria. Even a simple biotin deficiency may ruin an otherwise "good hair day" by robbing nourishment from the single muscle that holds up individual hairs.

Because of digestive enzymes and the strong concentration of hydrochloric acid, it is not surprising that there are few bacteria, friendly or unfriendly, able to stay in the stomach for long. As we age, and the amount of the hydrochloric acid in the stomach diminishes, some unfriendly bacteria can and do invade the GI tract. Babies who are not breastfed will probably not develop the presence of the very valuable and friendly *Bifidobacterium infantis*. In addition, adults who lead a stressful life or take antibiotics cannot expect to have a high population of *Lactobacillus acidophilus*. As friendly bacteria that produce a useful antibiotic named acidophillin, this important member of our microflora helps to eliminate the presence of a wide variety of dangerous toxin-producing bacteria. Because *L. acidophilus* is an aerobic bacteria, it uses up the available oxygen to create an anaerobic environment that works in favor of certain friendly bacteria (*bifidus*) and against many of the toxic ones. Also *acidophilus* produces several short-chain fatty acids (e.g., lactic acid, butyric acid), which destroy the very harmful *E. coli* bacteria and keep the intestines healthy by keeping the pH low. *L. acidophilus* produces a variety of other helpful antibiotics (e.g., acidolin, lactocidin, lactobacillin) and even hydrogen peroxide, all of which will neutralize the dangerous effects of food poisoning from salmonella and shigella dysentery.

Putrefactive unfriendly bacteria (e.g., bacteroides) seem to increase their toxin production when *acidophilus*, and especially *bifidus* bacteria, are low. Meat eaters who have problems with bowel regularity are especially at risk. Phenols, indoles, and many amine waste toxins have been implicated in contributing toward colon cancer. Unfriendly intestinal bacteria also secrete toxin-creating enzymes such as dehydroxylases and glucuronidases, which cause cancer or other enzymes that can change cholesterol into coprostanol and coprostanone, which are known carcinogens.

By some estimates, 500,000 infant deaths worldwide per year might be prevented by reducing the population of *Clostridium fragilus*, the unfriendly bacteria responsible for diarrhea. Indeed, a variety of auto-immune diseases may also occur from excessive unfriendly bacteria.

As antibiotic usage, stress, and drug use increases, we will also see a decrease in friendly bacteria with a very predictable rise in yeast overgrowth from *Candida albicans*, enterocolitus from *Clostridium dificile*, and many other digestive and allergenic problems such as heartburn, constipation, and diarrhea.

Candida albicans

Candida albicans is a common yeast present in almost every human to varying degrees. It lives on the surface of mucous membranes. *Candida* often causes PMS, food sensitivities, hives, depression, endocrine disturbances, prostate pain, psoriasis, and even carbohydrate cravings! A variety of allergic reactions often originate in an intestinal tract overrun by *Candida albicans*. As *Candida* continues to proliferate, the immune system further weakens and opens up the possibilities of even more dangerous problems such as irritable bowel syndrome. Definitive diagnosis usually involves a stool culture or the careful measurement of elevated antibody levels to *Candida*.

The problem of yeast overgrowth may be brought on for a variety of reasons and by a number of stress factors. Antibiotics such as tetracycline, corticosteroids, anti-ulcer steroids, oral contraceptives, too much sugar, a lack of digestive secretions, or even an Epstein-Barr viral infection are some of the more common causes for yeast overgrowth. Whenever the balanced ecology of our intestinal microflora is disturbed by such stress factors, *Candida* will often proliferate to the point of challenging and compromising the status of our very immune system.[6] Once again our friendly bacteria, especially *L. acidophilus*, produce the right biochemical environment to push back the onslaught of a *Candida* infection. Aside from "friendly antibiotics," biotin is biosynthesized, which helps to prevent the *Candida* yeast from advancing to the even more detrimental fungus stage. At that point, *Candida albicans* can grow mycelial roots (rhizomes) that can penetrate the GI lining, causing "leaky-gut" syndrome. A variety of allergies are then free to plague us as our immune system continues to produce antibodies for the food antigens that have entered the bloodstream.

Chronic Infections

Once the ecology of our intestinal microflora balance is altered and disrupted, a general condition of "bowel dysfunction" sets in, opening the way to a parade of possible chronic infections. *Candida* and other unfriendly bacterial overgrowth begin to take over the health of the intestines. Invasion of tissues takes place all along the GI tract,[7] as well as inside the liver, bladder, lungs, and prostate.[8] Even the heart and thyroid glands are susceptible to *Candida*. In some infections, *Candida* can sometimes invade any and all human tissue cells by traveling through the bloodstream. In this manner, *Candida* becomes resistant to conventional therapies, especially the use of topical medicines.

The daily consumption of yogurt with *Lactobacillus acidophilus* has decreased the number of yeast infections in women. Unfortunately, most yogurt does not contain enough live bacteria to be helpful.[9] For some, a yeast-free diet may be needed. Bread, vinegar, grapes, ketchup, sugar, alcohol, peanuts, milk products, dried fruit, melons, potatoes, corn, and pickles should be avoided. Vitamins A, B_6, and folic acid, along with zinc, selenium, magnesium, and essential fatty acids should be emphasized in the diet.[10] Daily consumption of AFA algae is very helpful to supply these *Candida*-destroying nutrients.

As *acidophilus*, essential fatty acids, and digestive juices such as HCl, bile, and pancreatic enzymes increase, *Candida* population substantially begins to decrease.[11] In addition, garlic may be used to kill unfriendly bacteria. Its use may be more potent than the often prescribed medicine "Nystatin" to rid the GI tract of unfriendly bacteria.[12] Sometimes goldenseal, German chamomile, or ginger have helped clean up the GI tract as well. All plants, in general, produce chemicals that protect themselves from microbial invasion. There are compounds (such as phenyl isothiocyanate in soy beans) in many vegetables that are strong enough, when ingested by humans, to be able to kill *Candida*.

One of the best ways to protect ourselves from *Candida* invasion is to eliminate refined sugars, which only accelerate the growth rate of this yeast.[13] It is understandable how the high glucose levels of diabetics compounds their disease with *Candida* overgrowth. Human breast milk may contain a "growth-enhancer factor," as yet unknown, that works well with the *L. acidophilus* present in the milk. AFA shares some compositional similarities to human breast milk, partly because of its rich array of amino acids and essential fatty acids. Because AFA has been shown earlier to strengthen the immune system (as does human mother's milk), AFA algae and probiotics combine to make a superb team to combat the infections of *Candida*.[14]

Many people, after ingesting AFA with a probiotics regimen, begin to experience cleansing symptoms. This response is normal. It is your body's way of discharging toxins, some of which may have been lodged in your cells for a very long time. Thus, you might now reach a new level of health. This is also a good time to emphasize fresh fruits and vegetables, as well as whole grains in your daily diet.

Candida Toxins

As *Candida* spreads throughout the body, toxic compounds (e.g., ethyl alcohol, acetaldehyde, and tyramine) are produced. Certain strains of *Candida* can overrun the GI tract and actually produce enough ethyl alcohol (i.e., drinking alcohol) to cause mild intoxication![15] This overgrowth of *Candida* also brings about a variety of common mood swings, headaches, and even chronic depression. These symptoms are most likely due to the inability of the liver to completely detoxify enough of the *Candida* by-products. In addition, the liver can eventually be overwhelmed with the toxins created by dead yeast cells as well. Because of this, the symptoms of candidiasis will often noticeably worsen.[16]

Typical Symptoms of Bowel Dysfunction from Candida

- CANKER SORES in the mouth may indicate either a sensitivity to certain foods or a proliferation of *Candida* at the expense of friendly microflora.
- HEARTBURN AND HIATAL HERNIA are often diagnosed along with canker sores.[17]
- GAS, BLOATING, INDIGESTION, AND NAUSEA are sometimes brought on by the release of carbon dioxide gas due to the proliferation of *Candida.*
- DIARRHEA is very common when *Candida* has taken over the colon and caused either an infection or an allergic response.[18]
- CONSTIPATION symptoms can be greatly reduced or prevented with a proper probiotic regime. Unfortunately, there are other causes of constipation of unknown origin.[19]
- RECTAL AND VAGINAL ITCHING are often caused by a dysfunctional bowel or other unknown allergic reactions.
- NASAL CONGESTION AND CHRONIC SINUSITIS may occur after *Candida* proliferation impairs the immune system such that a *Staphylococcus* infection may continually manifest.
- HEART POUNDING AND PALPITATIONS are remarkably common complaints in people with candidiasis. These and other anxiety symptoms can be sometimes cleared up with a probiotics regimen.[20]

Probiotic Benefits in Fighting Back Candida

There are numerous benefits in taking probiotics:

- Probiotics improve the immune system by preventing the overgrowth of *Candida.* This is because *Candida* is known to greatly impair the entire immune system by decreasing the number of T-cells and natural killer cells.[23]
- *Lactobacillus* probiotics manufacture B vitamins, vitamin K, and even digestive enzymes.[22] Thus friendly bacteria can prevent deficiencies of certain vitamins.

- Low-density lipids (LDLs) carry cholesterol to our cells and are thus often called "bad cholesterol." Some researchers believe that friendly bacteria actually consume cholesterol and thus reduce the incidence of coronary diseases.
- Dangerous nitrites from packaged meats and other such foods are known to be converted to cancer-causing nitroso-compounds by the enzymatic action of various unfriendly bacteria. The *Lactobacillus* bacteria in probiotics seem to be able to metabolize and render such compounds relatively harmless.
- If rampant overgrowth of *E. coli* is not held down, there is strong evidence that a diabetic reaction may result. Apparently, the unfriendly *E. coli* produces a toxin that mimics insulin, thus blocking real insulin receptor sites on cells that need its sugar-regulating effects. Immediate reduction of fats and sugars (along with probiotic therapy) is recommended. A strong population of *bifidum* bacteria keeps E. coli from spreading.
- An overgrowth of the unfriendly *Yersinia enterocolitica* may lead to endotoxin production, which seems to stimulate overactivity of the thyroid. These bacteria are combatted by probiotics.
- Too much meat eating, along with a dysfunctional bowel with unfriendly bacterial build-up, probably creates excessive ammonia gas production that invariably puts a stress on the liver to keep up with its detoxification. A good supplementation with bifidobacteria (supplemented with AFA) will greatly reduce such stress on the liver.
- Leaky-gut syndrome can allow bacterial debris and other toxins (phenols, ammonia) to enter the bloodstream and cause an antigenic white blood cell response. This will put the body into a constant state of "disease alert," which can manifest itself in migraines and learning disorders. Over a period of several months, a wide range of friendly microflora can often repair the problems of a toxic bowel. These micro-companions are:

    1. *Lactobacillus acidophilus*
    2. *Lactobacillus bulgaricus*
    3. *Lactobacillus plantarum*
    4. *Lactobacillus salivarius*
    5. *Bifidobacterium bifidum*
    6. *Enterococcus faecium*
    7. *Spectrococcus thermophilus*

The four *Lactobacillus* species listed above provide us with absorbable B vitamins and help us to biosynthesize several digestive enzymes. If these microflora are depleted and *Candida* is allowed to proliferate and cause inflammation and diarrhea, nutritional status will be impaired partly because *Candida* will actually greedily utilize the B vitamins and other nutrients for its own growth.[24] Loss of the digestive enzymes produced by *Lactobacillus* bacteria also reduces the amount of assimilated nutrients.

A wide body of experimental evidence has recently been published in medical journals that shows that a variety of endocrine gland problems (possibly manifesting in some cases as hypothyroidism or hyperthyroidism, premenstrual stress syndrome, or type 1 diabetes) are connected with the proliferation of unfriendly intestinal microorganisms.[25]

- Probiotics may help reduce the uncomfortable sensitivity to foods and chemical odors.[21] This is because Candida toxins, specifically glycoproteins that cause histamine release, are known to cause a variety of allergic reactions.

WITH THE HELP OF FRIENDLY MICROFLORA such as bifidus and acidophilus, we can enhance our health by restoring the natural ecology of our gastrointestinal tract. They help to selectively destroy the toxic microflora which lurk in the deep recesses of the small and large intestines and prevent them from secreting carcinogenic waste toxins. Thus AFA and probiotics, along with the digestive enzymes discussed next, create a unique health-promoting team.

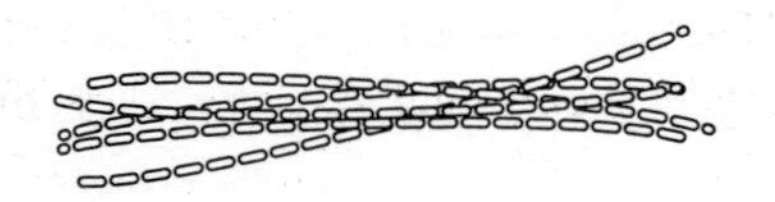

CHAPTER FOURTEEN

P.S. – Don't Forget Your Enzymes

CENTRAL TO THE DIGESTION OF FOOD IS THE DISCIPLINE OF careful chewing. Although many of us are aware that enzymes in the mouth are required to break down complex carbohydrates to digestible sugars, most of us are not aware of the crucial role that chewing our food plays in the release of enzymes when the cells of raw fruits and vegetables are ruptured.

Apparently, all animals have a digestive system that allows time for digestive enzymes to take effect in the ingested food. For example, when birds ingest grains or seeds, the protein- and starch-digesting enzymes within the grains and seeds act for hours inside the bird, promoting the eventual digestion and assimilation of the nutrients within. Of course, most of us do not eat raw and uncooked vegetables or grains. We prefer, for good reason, to cook such food. It is easier to chew and often tastes better. Besides, most of us have, in this fast-paced world, long since given up the patient process of chewing our food properly so that all endogenous enzymes will be released.

All animals, including humans, predigest their food in a separate, upper portion of the stomach sometimes called the *food-enzyme stomach*. The upper portion of the human stomach is this reservoir wherein amylase (from the ptyalin of the saliva) and other exogenous food enzymes (from the food or from supplements) carry out their digestive actions without muscular peristaltic motion. This, of course, implies that the human stomach is naturally designed to allow food the necessary time to slowly digest within the food's own internal enzymes. But since cooked foods no longer contain their own internal enzymes, the food-enzyme stomach is robbed of most of its digestive purposes. This puts a tremendous strain on other digestive organs, enlarges the liver and the pancreas, and sets the stage for a dark parade of modern degenerative diseases that work invisibly over many years, manifesting without warning as a breakdown of one or another exhausted organ. Most laboratory researchers and technicians are aware of the sad fact that laboratory animals, especially rodents, develop kidney disease from the enzyme-free diet they are typically given. Fortunately for humans, exogenous digestive enzymes can begin to reverse these effects on ill-advised diets of processed foods.

Endogenous and Exogenous Digestive Enzymes

When food is cooked, its endogenous enzymes are destroyed (denatured). Thus, unless a fermented food (i.e.., pickles, miso, sauerkraut, yogurt, etc.) is included to digest cooked foods, our digestive system must compensate by producing more of the needed and appropriate digestive enzymes. Saliva contains amylase, which digests carbohydrates; and stomach juices contain proteases, which digest proteins. Later, the pancreas secretes lipase into the intestine to digest fat, along with additional amylases and proteases. Chewed and swallowed food settles first in the upper region of the stomach for 30 to 60 minutes wherein endogenous salivary enzymes (amylase) continue to digest carbohydrates (leaving proteins and fats for later) and exogenous enzymes from raw food begin to digest other food groups. Predigestion occurs in the upper portion of the stomach until gastric acid secretions begin to inactivate the endogenous digestive enzymes.

Cooked food is devoid of digestive enzymes and therefore stimulates our internal enzyme secretion. Raw food is rich in digestive enzymes and only stimulates a small amount of endogenous enzyme secretion. Since less stomach acid is secreted, raw food remains longer and digests better in the upper stomach.

All uncooked foods are known to contain enough endogenous enzymes to digest the foods contained within. That is, *enzymes digest their own foods*. Raw foods, for example, that are high in fats are equally high in lipase. Raw honey and bananas contain enough amylase to transform their carbohydrates into simple sugars. Even raw meat contains enough endogenous cathepsin protease to easily digest its own protein content. Some meat tenderizers commonly used today use enzymes from unripe papaya or other sources similar to the enzyme-rich fungus *Aspergillus*.

Surprisingly, unprocessed raw milk has been found to contain more than 20 digestive enzymes, with lipase enzymes being the most important.[1] Before the advent of pasteurization, raw milk was even used in some enzyme therapy programs dating all the way back to the time of Hippocrates. Apparently, the lipase content reverses the artery-clogging effects of cholesterol and atherosclerosis.

The Art of Medicine consists of amusing the patient while Nature cures the disease.

VOLTAIRE

The raw fish and animal fat eaten by Eskimos contains enough endogenous enzymes (cathepsin and lipase) to protect them from heart disease. In their own language, the word "Eskimo" actually means "he who eats raw food."

The gluten within wheat consists of a single polypeptide (mostly glutamine and proline) that can produce an abnormal immune response[2] and sometimes brings on celiac disease. Fortunately for bread lovers, the enzyme papain in papaya may be used as a supplemental enzyme to help render wheat gluten harmless.[3]

After years of digesting cooked foods that no longer contain any available enzymes of their own, physiological stress begins to take its toll on all of us, and degenerative diseases begin to show the early symptoms of pathological enlargement of regulatory glands such as the pancreas and the pituitary. Since processed foods require more pancreatic enzymes for their digestion, animals and humans that eat such enzyme-depleted foods are consistently found to have a correspondingly greatly enlarged pancreas (200 to 300 percent increase).

Enzymes Bioavailable in AFA Algae

As we have seen earlier, enzymes are biocatalysts that speed up reactions in the body, be they digestive or metabolic. For example, digestive enzymes called proteases are able to break down and digest protein molecules. Reactions such as these occur only when the enzyme acts on that molecule's shape so that its bonds break just as they should.

The cellular activity of AFA depends upon hundreds of such specific enzyme reactions. Being among the most independent of all unicellular organisms, AFA is able to efficiently use its specialized enzymes to change the basic ingredients of earth, air, water, and sunlight into the molecules needed to build its body and maintain its life.

Superoxide dismutase is an example of an enzyme that AFA uses to protect its DNA from dangerous "superoxide" free radicals. As cited earlier, superoxide dismutase (SOD) requires copper, zinc, and manganese as mineral cofactors. By ingesting and digesting AFA's enzymes, we are being nourished with an array of its chelated vitamins and minerals, as well as the amino acids that compose it. As a result, AFA has a stimulating effect on the biosynthesis of many other enzymes in the body.

Enzymes – whether endogenous or exogenous – are either metabolic or digestive in purpose. There are thousands of different metabolic enzymes at work in the human body; almost a hundred metabolic enzymes work within the human arteries alone. There are only about 21 digestive enzymes, classified as proteases (which digest proteins), amylases (which digest carbohydrates), or lipases (which digest fats).

Important life-sustaining chemical reactions can take place only in the presence of highly specialized enzymes, which remain bioactivated because they have not lost – in the words of some researchers – their *vital life force*. Enzyme bioactivation and consequent vital life force depend not only upon the enzyme's mineral, vitamin, and amino acid content, but also upon the temperature and acidity of its environment. The 1946 Nobel Prize in biochemistry was awarded to James B. Sumner for his pioneering work in the crystallization of such enzymes. According to *Science News*,[4] scientists are still studying how enzymes control the products of chemical reactions.

Every enzyme, and there are thousands of them, carries out a very specific reaction. Each of our digestive enzymes specifically acts upon a different food type. Being deficient in one or missing the full activity of another can mean sickness or even death. Pancreatic enzymes also keep the small intestine free from parasites (yeast, bacteria, intestinal worms) and break down immune complexes.[5] Vibrant health may be defined as having all of our enzymes working together like a vast symphony.

Because the life force energy of exogenous digestive enzymes found in raw foods is destroyed within processed foods by their cooking, pasteurization, and now irradiation, too many endogenous digestive enzymes are expended, in turn depleting the body's reserve of metabolic enzymes. Ingesting more exogenous digestive enzymes reduces the substantial loss of endogenous digestive enzymes needed during food consumption. This increases the overall quantity of enzyme reserve or "enzyme potential" of endogenous metabolic and digestive enzymes and greatly enhances the health of the individual. The body now has more metabolic enzymes in its reserve to build, repair, and protect itself. Death results when the enzyme reserve is depleted.

Plant-Derived Oral Digestive Enzymes

Rather then giving up cooked food, our solution is relatively simple: before each meal of cooked food it is strongly recommended that plant-derived enzymes be taken orally so that some of this digestive stress is circumvented. Plant enzymes derived from the fungus *Aspergillus oryae* seem to be particularly effective. For almost 50 years, *Aspergillus* has been used in Japan to enzymatically ferment soybeans. These digestive enzymes are particularly good for digestive stress reduction because they are able to function in spite of the wide range in pH throughout the body, from the stomach to the small intestine. The proteolytic enzymes in *Aspergillus* contain amylase, protease, lactase, lipase, and cellulase.

Some of the plant enzymes from Aspergillus that are not used up while assisting digestion are able to pass through the intestinal mucosa and enter the bloodstream. Once there, they can be helpful by hydrolyzing (chemically digesting) circulating proteins that the immune system otherwise considers to be invaders. This takes an enormous strain off the immune system, allowing it to deal with the other invaders that do pose a real threat.

Improving Protein Absorption

There is strong evidence that the normal digestive process allows about 2 percent of dietary protein to be absorbed across the intestinal wall into the bloodstream, causing an antibody reaction and food allergies.[6] Fortunately, the normal gastrointestinal tract contains large enough quantities of immunoglobulin A (IgA) proteins to bind to undigested proteins and other dietary toxins before they enter the bloodstream.

Poor intestinal absorption of proteins and amino acids leads to their bacterial breakdown and consequent production of foul-smelling polyamine bowel toxins such as putrescine. This leads to bowel toxemia, which quickly begins to impair the function of the liver. Since the liver essentially filters the blood, excessive toxins will often lead to psoriasis. This is one reason why protease supplementation from Aspergillus is so strongly recommended.

Circulating Immune Complexes in Need of Enzymes

Food allergies are sometimes caused by undigested dietary proteins that appear as foreign invaders after they have leaked through the intestinal mucosa into the bloodstream. This is caused by the absence or dysfunction of protein-digesting enzymes. By taking oral *Aspergillus* as a digestive enzyme supplement, such antigen-antibody "circulating immune complexes" (CICs) can be more easily and enzymatically broken down and digested. Incompletely digested foods lead to "digestive leukocytosis" because such particulate matter is attacked in the blood as a foreign antigen. Blood loss from the gut[7] often brings on iron deficiency. This is especially true for those with Crohn's disease. Science still needs to find ways to improve our iron absorption because too much left in the gut can actually promote intestinal infection by feeding unfriendly microorganisms.[8]

A majority of children with asthma have low stomach acidity. Because of this they can often even develop new allergies.

Proteolytic, or protein-digesting enzymes, such as pineapple bromelain or *Aspergillus* protease are effective in reducing the inflammation brought on by food sensitivities.[9] Such proteases reduce swelling by allowing the body to break down the CICs that often concentrate in our joints. In general, taking enzymes before meals is helpful for food digestion; taking enzymes between meals is helpful for combatting allergies.

Undigested proteins can cross the intestinal barrier and become absorbed into the bloodstream, causing a variety of food allergy responses.[10] This may be brought on by low HCl and insufficient digestive enzyme secretion, as well as a poor immune system, low betacarotene, and abnormal proliferation of unfriendly bacteria.[11]

Digestive enzymes circulating in the bloodstream eventually find their way back to the pancreas. Once there, they are able to be reused to some extent by pancreatic cells. This is analogous to the bile salts that are recycled and re-secreted by liver cells.

Chewing exposes food to hydrochloric acid (HCl) and enzymes. Natural endogenous raw food enzymes are bioactivated by moisture and exposure to oxygen in the air provided by the chewing process. Betaine HCl is needed in protein digestion. Many menopausal women are severely deficient in HCl. Rheumatoid arthritics usually have a low production of HCl and a high sensitivity to certain protein foods.[12] Sometimes low secretion of digestive juices and enzymes can even disturb our delicate balance of microflora.

Healing Properties of Oral Enzyme Therapy

THERE HAS BEEN A SIGNIFICANT AMOUNT OF RESEARCH in Europe on the use of oral plant-derived digestive enzyme therapy. Much of the research was pioneered by Henry Lindlahr, Francis Pottenger, and particularly by Edward Howell.

Wild animals appear to be immune from the hundreds of new and modern diseases of man that are brought on by diets almost devoid of bioactivated enzymes, although pollution and the pesticiding of America has certainly wielded catastrophic damage to so many. Chronic pancreatitis is associated with the common cause of insufficient secretion of pancreatic digestive enzymes. Failure to digest fats leads to malabsorption and severely compromised nutritional health. Exogenous lipase, the enzyme needed to digest lipids, can be derived from *Aspergillus* and has been shown to be highly effective in reversing such conditions. This enzyme has also been shown to be effective in treating and preventing the vascular diseases that result from obstructed arteries. Even some forms of adult acne (rosacea) may be caused by a gastrointestinal disorder involving too little pancreatic lipase output.[13]

Medical researchers have demonstrated how a variety of circulating immune complexes were successfully treated; even immune complexes in the kidney were dissolved and broken up by oral enzyme treatment.

Lactose intolerance is a condition that mostly affects Africans (more than 75 percent), Asians, and Mediterranean adults. For people with that intestinal disorder, lactase, the disaccharide enzyme that digests the lactose sugar in milk products, is deficient. Affected children lose this enzyme after age 3 to 7. Oftentimes, after an acute viral or bacterial illness injures the small intestine's mucosal cells, there will even be a temporary lactase deficiency.[14]

Common symptoms of lactose intolerance are abdominal pain, cramping, flatulence, and diarrhea. Milk products should, of course, be avoided by those with lactose intolerance. *Aspergillus*-derived lactase is one good source of the enzyme that is effective in treating lactose intolerance. This condition can also be helped by ingesting *acidophilus* bacteria and allowing them to manufacture some of this important enzyme.[15]

Since enzymes eventually wear out and are excreted by the body, their constituent vitamins, minerals, and amino acids need to be continually replaced by wholesome enzyme-activating foods. Digestive enzymes in humans under age 30 are roughly from two to ten times more active than in humans over age 60. The aging process itself may be directly related to the activity and strength of all endogenous enzymes. Bromelain is a proteolytic enzyme found in pineapple. Clinical studies demonstrate that if bromelain is taken between meals, there is a considerable reduction in pain and faster healing time for bursitis and tendonitis.

In general, raw foods are enzyme-rich and life-supporting. AFA is enzyme-rich and, because of its abundant supply of natural vitamins, minerals, and essential amino and fatty acids, it is also *enzyme-activating.*

BIOSTIMULATED BY SUCH A PROFOUND RANGE OF AFA *micronutrients, every cell in our body is enzymatically encouraged to become as flexible and efficient as it can be. The intelligence of the AFA can then be better translated to the intelligence of our own cellular world.*

Inspired by nutritional blue-green algae,

the body begins to

heal itself.

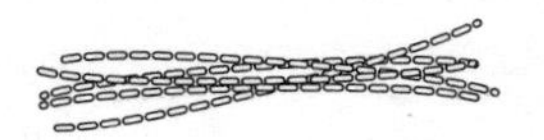

Epilogue

IT'S 12:30 IN THE MORNING, AND TOMORROW THIS MANUSCRIPT goes to the printer. So this is the last opportunity (for now) that I'll have to tell you how I feel. Not about formulas and studies or biostimulating chemical reactions – but to express, without the use of books or charts or research papers, that it's now time to pass the baton to you, the reader.

Enough has been said about nutritional and physical health issues. Now we must concern ourselves with the next phase of our development. As an ever-growing family of peace-seeking individuals, we must reach out to those who will benefit most from a regained sense of purpose. Happily, the human being is heroically much more than just a miraculous collection of organs and systems laboring non-stop to maintain a life. Almost unfathomable is the true meaning of heart, faith, desire, will, and soul. My overwhelming desire now, is to witness, one by one, men and women helping the elderly achieve some serenity and the next generation to inherit some hope.

Let's never stop trying to create a cleaner world, rich in harmony and capable of caring.

– K.J.A

APPENDIX A

Lipid A – Focus of Discovery

A Special Immume System Booster

EMBEDDED WITHIN THE OUTER LAYER OF THE CELL WALL of certain types of bacteria and cyanobacteria (blue-green algae) is an interesting type of lipid molecule – sometimes called Lipid A – that has become an exciting focus of attention for many scientists all over the world.

The exact shape and concentration of these specialized cell wall lipids characterize the species of one or another single-celled organism. These lipids consist of variable fatty acids and phosphate groups that are found attached to sugar chains (polysaccharides) on the cell wall surface. Such lipopolysaccharides (LPS) range from extremely low (homeopathic) amounts for some species like AFA, to the much higher concentrations found in certain varieties of toxic bacteria such as *Salmonella*. Typically, the lipid portions of the LPS molecules directly stimulate our white blood cells in different ways depending on the structure and amount of the LPS. Subtle white blood cell stimulation by the lipids of AFA causes immunostimulating responses that benefit the entire immune system.

This benefit is a direct result of our white blood cells' response to AFA's small amount of cell wall lipids. By releasing special proteins (interleukins and "tumor necrosis factor"), our white blood cells can better defend us against microbial invaders and cancer threats.[1] In some cases, these mysterious immune-boosting proteins travel through the bloodstream to the temperature control center of the brain (hypothalamus) and may also induce cold-like symptoms. This is the detoxification process occasionally experienced by some algae eaters.

Currently, biochemists and biologists are pursuing the discovery that tiny fragmentary pieces of LPS can also block receptor sites on white blood cells. Because of this, detox responses begin to diminish and a new plateau of immune system health is approached.[2]

APPENDIX B

Noteworthy News Articles

ACCORDING TO A RECENT AND REVOLUTIONARY FEDERAL study, environmental toxins are even worse than previously thought. While individual pesticides are dangerous even in small quantities, two or more pesticides entering our body in the same day will have toxic affects hundreds or even thousands of times worse than previously believed! Breast cancer and male birth defects are particularly affected. Pesticides have these affects on our body because of their disruptive effect on portions of our DNA that manufacture estrogen. Dr. Lynn Goldman of the EPA Office of Prevention, Pesticides and Toxic Substances told the Associated Press (June 7, 1996) that this alarming realization is now "very high priority" and new chemical testing of our foods are now necessary because this negatively synergistic impact of multiple toxins "is a very, very new issue for us."

UNFORTUNATELY, EVEN THE TINIEST TRACES OF ENVIRONMENTAL toxins can seriously compromise the immune system. According to Robert Hotz, Science Writer for the Los Angeles Times (May 31, 1996), our children are in the greatest danger because of the "profound and irreversible damages such toxins cause to an infant's memory and learning abilities." Human sensitivity to such toxins is high because these pesticides were *intentionally* designed by chemists to mimic hormones. Even small amounts in a pregnant woman can have disastrous affects on the future of the fetus.

Until we live in a toxin-free environment, and until our immune systems are no longer compromised from the impact of multiple pesticide affects on our body, we need to protect ourselves as best we can. With world-wide pollution so prevalent, fortuitously AFA algae has come to our rescue!

IT HAS RECENTLY BEEN SHOWN THAT BLUE-GREEN ALGAE CAN significantly prevent mouth cancer. Padmanabhan Nair at Maryland's Human Nutrition Center administered only one gram a day to tobacco chewers in Kerala, India, who had precancerous mouth lesions (oral leukoplakia). Dramatically, after one year, the sores had vanished or shrunk significantly in more than half of the people, with complete regression in 45 percent of the subjects. Dr. Nair speculates that the antioxidant pigments such as betacarotene help to remove damaging cancer-causing free radicals.[3]

IN JANUARY 1995, THE JOURNAL *CANCER RESEARCH* PUBLISHED a long-awaited study that finally confirms the intuitive belief that chlorophyll present in foods such as organic vegetables and AFA algae has anti-cancer properties! This article reports conclusively that chlorophyll is a "potent inhibitor of hepatocarcinogenesis ... and may have important implications in [the] intervention and dietary management of human cancer risks." [4]

SOME ALGAE, LIKE DUNALIELLA AND AFA, ARE KNOWN TO CONTAIN very high concentrations of natural betacarotene which serves as a very potent molecular shield, preventing lipid oxidation within our delicate cell membranes. Such algal betacarotene is naturally and equally divided up between "trans" (straight) and "cis" (bent) type variations (see illustration on page 27). Because both forms differ somewhat in molecular shape, their ability to absorb dangerous free radicals varies. In fact, according to the March 1994 Journal of Applied Physiology, it has been determined that the cis form protects cellular membranes more efficiently than the trans variation. However, both forms are needed for maximum protection. So powerful are their combined healing effects, people exposed to dangerous levels of oxygen-free radicals should be treated daily with algal carotenes.[5]

IN AN EXPERIMENT REPORTED IN the British *Journal of Cancer,* white pigs were given gamma linolenic acid (GLA) or a placebo for four weeks before and sixteen weeks after radiation exposure. Since the severity of skin reaction was significantly minified for the animals receiving GLA, British biologists concluded that GLA represented a safe way for helping to protect the skin from the ravaging effects of radiation.[6]

Algae show promise in cancer fight

BALTIMORE Blue-green algae, the pond scum that bedevils fishermen and is the bane of backyard swimming pools, has yielded a promising new family of anti-cancer drugs, researchers say.

Compounds derived from algae have shown remarkable ability to shrink tumors in mice, enough to cure some animals, said Gregory Patterson of the University of Hawaii.

Mice were implanted with cancer cells that cause breast and prostate cancer in humans and treated with algae-derived compounds called cryptophycins, which appear to attack the internal structure of cancer cells, blocking their ability to spread, Patterson said.

The research has led to an agreement with Eli Lilly & Co. of Indianapolis, which will develop the compound in hope of marketing an anti-cancer product for humans.

Compiled from Examiner wire reports

Friday, February 9, 1996 ★ ★ ★ ★

San Francisco Examiner

WHEN AFA ALGAE IS FREEZE-DRIED AND PRESSED INTO CAPSULES or tablets, exotic compounds within the cell wall are sometimes broken down into molecular fragments. Such fragments of lipopolysaccharide (LPS) or lipid A are known to have "immunopotentiating" effects. As reported in the *International Journal of Immunopharmacology,* such pieces of molecular cellular wall can greatly enhance the strength of the entire immune system.[7]

In the recently published March 1996 issue of the Scandinavian *Journal of Immunology*, PUFAs such as eicosapentaenoic acid (EPA) and especially docohexaenoic acid (DHA) are shown to be very potent inhibitors of "over-active" lymphocytes (white blood cells) that sometimes attack healthy tissues.[8] PUFAs, like those found in AFA and fish oil were shown to have been successful in improving the health status of people who suffer from autoimmune disorders.

❧

As reported in the *International Journal of Neuroscience*, there is a strong correlation between vitamin B_{12} deficiency and the possible onset of multiple sclerosis. B_{12} deficiency may "render the patient more vulnerable to viral mechanisms". Thus, the B_{12} in AFA is one very important and natural way to boost the immune system and protect the body against such viral invasions.[9]

❧

According to a study published in the *Journal of Internal Medicine* (June 1993), the human immunodeficiency virus (HIV) is known to be associated with dementia, difficulties in concentration, memory, and motor function. When treated with vitamin B_{12}, 20 percent of the patients experienced relief of such symptoms within two months. It was concluded that such symptoms of dementia are indeed reversible for some HIV patients.[10] Even psychiatric disorders and related immune system abnormalities have been dramatically diminished in children with vitamin B_{12} supplementation.[11]

❧

It has been known for some time that the amino acid arginine can dramatically boost the immune system by increasing the production of natural killer cells that are capable of inhibiting tumor growth.[12] A 1994 double-blind, placebo-controlled clinical trial reported in the journal *Cytokine*, describes how low-dose dietary arginine supplementation in 29 patients with diabetes mellitus brought about a significant two-fold enhancement of the immune system.

This is especially interesting because AFA algae contains copious chains of interconnected arginine molecules generously and safely stored inside unique granules deep inside the cell.

AFA, LIKE HUMAN BREAST MILK, CONTAINS A PEPTIDE KNOWN AS substance P, an immune system booster. According to a 1995 article in the *American Journal of Clinical Nutrition,* the absence of substance P in infant formula may seriously impair the nutritional status of the developing infant.[13]

A 1994 STUDY PUBLISHED IN THE *JOURNAL OF NUTRITIONAL Science and Vitaminology* showed that when laboratory mice were fed a blue-green algae diet, there was a marked improvement in the overall ability of the spleen to mount an effective defensive immune response. This was accompanied by a significant increase in the ability of white blood cells to engulf bacterial and viral invaders through phagocytic activity with intensified interleukin production.[14]

APPENDIX C

AFA & Spirulina Compared

FOR THE PAST SEVERAL DECADES, PEOPLE HAVE BEEN ENTHRALLED by a "green foods" revolution. During this time, several foods have been championed as the revolution's leader. Foods such as barley grass, *Chlorella,* wheat grass juice, and sprouts lag far behind the two most popular blue-green algae, *aphanizomenon flos-aquae* (AFA) and *Spirulina*. Both are considered green superfoods;

they have similarities but several important differences. The subtleties of their differences have become evident more recently, as more sophisticated methods of analysis are employed.

One major difference is simply that AFA is the "greenest" superfood known, because it has the most of that wonderous green photosynthesizing pigment chlorophyll. A ten-gram portion of AFA algae contains 300 mg of chlorophyll, whereas a ten-gram portion of *Spirulina* has only 115 mg.

Both forms of blue-green algae have a similar overall concentration of proteins, carbohydrates, lipids, and minerals. However, the quality of their micronutrients is noticeably different because of specific growing and harvesting techniques associated with them. *Spirulina,* for example, is grown in concrete or plastic ponds with a "salinity factor" (sodium chloride salt content) often greater than 100 times that of AFA algae. *Spirulina*'s nutrient composition is just a reflection of the substances that have been artificially added to it in the form of mineral (and other) supplements. AFA, by comparison, is harvested from its own mineral-rich and natural Upper Klamath Lake habitat. Its micronutrients mirror what has existed naturally in the lake for thousands of years due to past volcanic activity and the interactions of rivers, streams, and unpolluted mountain rain, as well as a vast subterranean water supply originating from the nearby and pristine Crater Lake.

The richness of AFA's micronutrients are even more evident from the fact that 30 to 40 feet of organic nutrient sediment make up a treasure trove of minerals for AFA to feed upon. The vast richness of this sediment is reflected in the fact that AFA has about 40 percent more calcium and 100 percent more chromium than does *Spirulina.* AFA algae even has approximately five to ten times the vitamin C content of *Spirulina.*

One interesting difference between these two cyanobacteria can be traced to the fact that whereas *Spirulina* is a tropical algae, AFA is a heartier cold-climate species. In the warmer climate of the tropics, the cell membrane of a *Spirulina* cell can easily maintain its flexibility by producing a rather high percent of *saturated* fatty acids. AFA algae, on the other hand, does not lead this life of "tropical luxury." The colder climate of Upper Klamath Lake forces AFA's cell membrane enzymes to compensate by ingeniously manufacturing specific poly-*unsaturated* fatty acids (PUFAs) that enhance its life-sustaining membrane flexibility.

To be sure, both forms of algae are blessed with a rich array of phytochemical antioxidants such as the carotenes. Although *Spirulina* contains slightly more betacarotene than AFA algae, one must be careful to take a more educated look. The presence of more PUFAs allows for wider variety of other carotenoids – such as alpha and gamma carotene – to be spread out within the cell membrane itself. The true healing power of betacarotene cannot be fully realized unless a variety of other structurally-related carotenoid compounds is present.

Carotenoid compounds in all forms of blue-green algae are also particularly sensitive to the type of harvesting techniques employed. The sun-drying and spray-drying techniques often used in processing *Spirulina* invariably cause a marked decrease in betacarotene as well as the concentration of methionine – a sulfur-containing essential amino acid.

The assimilation of algal protein is also dependent upon how it is processed. When *Spirulina* is sun-dried or spray-dried, its "net protein utilization" (usually expressed as *percent assimilation*) is typically half that of AFA algae, which has been more carefully freeze-dried and flash-frozen. In general, freeze-drying techniques are essential to maintain the viability of AFA enzymes and its delicately-chelated minerals and vitamins. Spray- or sun-dried *Spirulina* products tend to readily lose much of such heat-sensitive components.

Endnotes

Chapter One

1. Kay, R.A., "Microalgae as food supplement," *Clinical Reviews in Food Source and Nutrition,* Vol. 30, 1991, p. 566.
2. Doual, T.L., *UNESCO Courier,* May 1993, pp. 43-4.
3. Cover Story: Pennisi, Elizabeth, *Science News,* March 12, 1994.
4. Wallace, D.R., *Wilderness,* Winter, 1992, pp. 14-15.
5. Carmichael, W.W., "The Toxins of Cyanobacteria," *Scientific American,* Jan. 1994.
6. Wuthrich, B., "Adverse reactions to food additives," *Annals of Allergy,* 71(4) Oct., 1993, pp. 379-84.
7. Cooper, K.H., 1994. "Antioxidant Revolution," 1st Ed. Thomas Nelson Pubs.

8. Kay, R.A.

9. Murray, M., and Pizzorno, J., 1991, *Encyclopedia of Natural Medicine,* Prima Pub., Rocklin, Calif.

10. Tobi, M., Morag, A., Ravid, Z., et al., "Prolonged atypical illness associated with serological evidence of persistent Epstein-Barr virus infection," Lancet, 1982, I, pp. 61-4.

11. Braverman, E.R., 1987. *The Healing Nutrients Within: Facts, Findings and New Research on Amino Acids*, 1st Ed., Keats Pub., Inc.

12. Weiss, B., "Food Additives and Environmental chemicals as sources of childhood behavior disorders," J. Am. Acad. Child Psychiatry, 1982, 21, pp. 144-52.

13. Lester, M.R., "Sulfite sensitivity: significance in human health," Journal of the American College of Nutrition, 1995 June, 14 (3): 229-32.

Chapter Two

1. Sarkar, A., et al., "Betacarotene prevents lipid peroxidation and red blood cell damage in experimental hepatocarcinogenesis." Cancer Biochemistry Biophysics, 1995 Nov., 15(2):111-25.

2. Burton, G. and Ingold, K., "Betacarotene: An unusual type lipid antioxidant," *Science, 1984.*

3. Tinkler, J.H., et al., "Dietary carotenoids protect human cells from damage." Journal of Photochemistry and Photobiology, 1994 Dec. 26(3) :283-5.

4. Brock, T.D. et al., *Biology of Microorganisms*, 4th Ed., Prentice-Hall, Inc. 1984.

5. Murray, M., and Pizzorno, J., 1991, *Encyclopedia of Natural Medicine*, Prima Pub., Rocklin, Calif.

6. Odens, M., "Prolongation of the Life Span in Rats," Journal of the American Geriatrics Society. 21:450- 451, 1973.

7. Berthold, H.K., et al., "Evidence for incorporation of intact dietary pyrimidine (but not purine) nucleosides into hepatic RNA," Proceedings of the National Academy of Sciences of the U.S.A., 1995, Oct. 24, 92(22): 10123-7.

8. Kulkarni, A., et al. "Effect of Dietary Nucleotides on Response to Bacterial Infections." Journal of Par. Ent. Nutr. 10: 169-171, 1986.

Chapter Three

1. Brock, T.D., et al., *Biology of Microorganisms*, 4th Ed., Prentice-Hall, Inc. 1984.

2. Braverman, E.R., 1987. *The Healing Nutrients Within: Facts, Findings and New Research on Amino Acids,* 1st Ed., Keats Pub., Inc.
3. Dillon, J.C., et al, "Nutritional value of the alga spirulina." Plants in Human Nutrition, 1995, vol 77. pp. 32-46.
4. Saturday Evening Post. Nov.-Dec. 1995.
5. Braverman, E.R., 1987. *The Healing Nutrients Within: Facts, Findings and New Research on Amino Acids,* 1st Ed., Keats Pub., Inc.
6. Journal of the American Medical Association, Nov. 16, 1994.
7. Consumer Reports, Oct. 1995.
8. Braverman, E.R., 1987. *The Healing Nutrients Within: Facts, Findings and New Research on Amino Acids,* 1st Ed., Keats Pub., Inc.

Chapter Four

1. The Lancet, Jan 15, 1994.
2. Journal of the American Medical Association, May 11, 1994.
3. The Lancet, Oct. 3, 1992.
4. Grosch, W. and Laskawy, G., "Co-oxidation of carotenes requires one soybean lipoxygenase isoenzyme," Biochem. Biophys. Acta., 1979, 575, pp. 439-45.
5. Hill, E.G., Johnson, S. and Holman, R., "Intensification of essential fatty acid deficiency in the rat by dietary trans-fatty acids," J. Nutr., 1979, 109, pp. 1,759-67.
6. Ross, R. and Vogel, A., "The platelet derived growth factor," Cell, 1978, 14, pp. 203-10.
7. Gerrard, J.M., White, J.G., and Krivit, W., "Labile aggregation stimulating substance, free fatty acids and platelet aggregation," J. Lab. Clin. Med., 1976, 87, pp. 73-82.
8. Horrobin, D.F., et al., "The Nutritional regulation of T-lymphocyte function," Med. Hypothesis, 1979, 5, p. 969.
9. *Time* Magazine, Sept 5, 1994
10. Kremer, J., Michaelek, A.V., Lininger, L., et al., "Effects of manipulation of dietary fatty acids on clinical manifestation of rheumatiod arthritis," Lancet, 1985, I, pp. 184-7.
11. Bittiner, S.B., Tucker, W.F.G., Cartwright, I. and Bleehen, S.S., "A double-blind, randomized placebo-controlled trial of fish oil in psoriasis," Lancet, 1988, I. pp. 378-80.
12. Sander, T.A.B. and Roshanai, F., "The influence of different types of Omega-3 polyunsaturated fatty acids on blood lipids and platelet function in healthy volunteers," Clin. Sci., 1981, 64, pp. 91-99.
13. Horobin, D.F., Medical Hypothesis, 1979.

14. The New England Journal of Medicine, Jan 28, 1993.
15. Leake, A., Chisholm, G. and Habib, F., "The effect of zinc on the 5-alpha-reduction of testosterone by the hyperplastic human prostate gland," J. Steroid. Biochem., 1984, 20.
16. Scott, W.W., "The lipids of prostatic fluid, seminal plasma and enlarged prostate gland of man," J.Urol., 1945, 53.
17. Vegetarian Times, Oct – , 1994.
18. Horrobin, D.F. and Manku, M.S., "How do polyunsaturated fatty acids lower plasma cholesterol level?" Lipids, 1983, 18, pp. 558-62.
19. Ibid.
20. The Lancet, Jan 15, 1994.
21. The Lancet, Aug 21, 1993.
22. Muscle and Fitness, Jan 1995.
23. Univ. Calif. at Berkeley, Wellness Letter, June, 1994.
24. New England Journal of Medicine, July, 1994.
25. New England Journal of Medicine, Nov. 1994.
26. Terano, T., et al, "Eicosapentaenoic acid as a modulator of inflammation," Biochem. Pharmacol., 1986, 35, pp.779-85.
27. Ibid.
28. Sept. 2, 1995
29. The Lancet, June 10, 1995.
30. Brook, J.G., Linn, S. and Aviram, J., "Dietary soya lecithin decreases plasma triglyceride levels and inhibits collagen- and ADP-induced platelet aggregation," Biochem. Med. Metabol. Biol., 1986, 35, pp. 31-9.
31. January, 1993.
32. Rao, R., Rao, U. and Srikantia, S., "Effect of polyunsaturated vegetable oils on blood pressure in essential hypertension," Clin. Exp. Hypertension, 1981, 3, pp. 27- 38.
33. The Lancet, October 16, 1993.
34. Stone, K., Willis, A., Hart, M., et al., "The Metabolism of dihomo gamma-linolenic acid in man," Lipids, 1978, 14, pp. 174-80.
35. Bisgaard, H., "Leukotrienes and prostaglandins in asthma," Allergy, 1984, 69, pp. 413-20.
36. Ayres, S. and Mihan, R., "Acne vulgaris: therapy directed at pathophysiological defects," Cutis, 1981, 28, pp. 41-2.
37. Nutrition Action Newsletter; November, 1993.

38. Science News, March, 1994.

39. Cohen, Zvi, Norman. H.A., Heimer, Y.M., "Microalgae as a source of omega three fatty acids," Plants in Human Nutrition: 1995, vol 77., pg. 1-31.

Chapter Five

1. Journal of the American Medical Association; as reported by the Associated Press: Feb. 7, 1996.

2. Prevention, July, 1995.

3. Collip, P.J., Goldzier III, S., Weiss, N., et al., "Pyridoxine treatment of childhood asthma," Ann. Allergy, 1975, 35.

4. Lipton, M., et al., "Vitamins, Megavitamin therapy and the nervous system," Wurtman, R., et al. Nutrition and the Brain, Raven Press, New York, 1979.

5. J.A.M.A., March 2, 1994.

6. Chiodini, R.J., et al., "Possible role of microbacteria in inflammatory bowel disease," Dig. Dis. Sci., 1984, 29.

7. Prevention, March, 1995.

8. Botez, M., et al., "Neurologic disorders responsive to folic acid therapy," Can. Med Assoc. Journal,1976.

9. Spector, T. and Ferone, R., "Folic acid does not activate xanthineoxidase," J. Biol. Chem., 1984.

10. Lipton, M., et al.

11. The Lancet, August 12, 1995.

12. Science, June 10, 1994.

13. Tufts University Diet and Nutrition Letter.

14. Abalan, F., et al., "B_{12} deficiency in presenile dementia," Biol. Psychiatry,1985.

15. Kay, R.A., "Microalgae as food supplement," Clinical Reviews in Food Source and Nutrition, Vol. 30 (1991).

16. Turley, S., et al., "Role of ascorbic acid in the regulation of cholesterol metabolism and the pathogenesis of atherosclerosis," Atherosclerosis, 1976.

17. Krumdieck, C. and Butterworth, C.E., "Ascorbate, cholesterol, and lecithin interactions in atherosclerosis," Am. Journal of Clinical Nutrition, 1974.

18. Olusi, S.O., Ojutiku, O.O., Jessop, W.J.E., and Iboko, M.I., "Plasma and white blood cell ascorbic acid concentrations in patients with bronchial asthma," Clinica Chimica Acta, 1979.

19. Buck, M.G. and Zadunaisky, J.A., "Stimulation of ion transport by ascorbic acid through inhibition of 3":5"-cyclis-AMP phosphodiesterase in the corneal epithelium and other tissues," Biochemica Biophysica Acta, 1975.
20. New England Journal of Medicine, April, 1994.
21. Formmer, D.J., "The healing of gastric ulcers by zinc sulfate," Medical Journal of Australia, 1975.
22. Alexander, M., Newmark, H., Miller, R.G., "Oral betacarotene can increase the number of OKT4+ cells in human blood," Immunol. Letters,1985.
23. Grosch, W. and Laskawy, G., "Co-oxidation of carotenes requires one soybean lipoxygenase isoenzyme," Biochem. Biophys. Acta., 1979, 575, pp. 439-45.
24. Prevention, September, 1994.
25. Prevention, July, 1994.

Chapter Six

1. Brock, T.D., et al., 1984 Biology of Microorganisms, 4th Ed., Prentice-Hall, Inc.
2. July 29, 1995; Krieger, et al.
3. Medical World News; December, 1992.
4. Journal of the American Medical Association, April 20, 1994.
5. Bierenbaum, M.L., et al., "Long term human studies on the lipid effects of oral calcium," Lipids, 1972.
6. Halpern, S., 1979, Clinical Nutrition. Lippencott.
7. Anderson, R., USDA Human Nutrition Research Center
8. Mertz, W., "Trace minerals and atherosclerosis," Fed. Proc., 1982.
9. The Lancet, July 9, 1994
10. Morley, J.E., "Nutritional status of the elderly," Am. J. Med., 1986, 81.
11. Krause, M.V. and Mahan, K.L., Food, Nutrition and Diet Therapy, 7th Ed., W.B. Saunders, Philadelphia, PA, 1984.
12. Colgan, M., et al., "Do nutrient supplements and dietary changes affect learning and emotional reactions of children with learning difficulties? A controlled series of 16 cases," Nutr. Health, 1984.
13. Brunner, E.H., Delabroise, A.M., and Haddad, Z.H., "Effect of parenteral magnesium on pulmonary function, plasma cAMP, and histamine in bronchial asthma," J. Asthma, 1985, 22.

14. Hughs, A., et al., "Platelets, magnesium, and myocardial infarction," Lancet, 1965.
15. Rosenberg, I.H., et al., "Nutritional aspects of inflammatory bowel disease," Ann. Rev. Nutr., 1985.
16. Prevention, Feb. 1992.
17. Seelig, M.S., "Magnesium deficiency with phosphate and vitamin D excess: role in pediatric cardiovascular nutrition," Cardio. Medicine,1978.
18. Seelig, M.S.
19. Rosa, G.D., et al., "Regulation of superoxide dismutase activity by dietary manganese," Journal of Nutrition, 1980.
20. Muscle and Fitness, November, 1993.
21. Science News, May, 1995.
22. Canadian Reader"s Digest, Oct. 1992.
23. Editorial, "Problems with prescription drugs among elderly," American Family Physician, 1986.
24. Brock, T.D., et al., 1984 Biology of Microorganisms, 4th Ed. Prentice-Hall, Inc.
25. Nordstrom, J., "Trace mineral nutrition in the elderly," Am. J. Clin. Nutr., 1982.
26. Lancet, Sept 19, 1992.
27. Nordstrom, J., "Trace mineral nutrition in the elderly," Am. J. Clin. Nutr., 1982, 36.
28. Leake, A., et al., "Subcellular distribution of zinc in the benign and malignant human prostate: evidence for a direct zinc androgen interaction," Acta Endocrinol., 1984.
29. Shroeder, H., et al., "Cadmium Hypertension," Arch. Environ.

Chapter Seven

1. Sanchez, A., Reeser, J., Lau, H., et al., "Role of sugars in human neutrophilic phagocytosis ," Am.J.Clin.Nutr., 1973 26, pp. 1,180-4.
2. Ibid.
3. Viti, A. et al., "Effect of exercise on plasma interferon levels," J. Appl. Phys., 1985.
4. Dillon, K.M., Minchoff, B. and Baker, K.H., "Positive emotional states and enhancement of the immune system," Int. J. Psychiatry Med., 1985-6, 15, pp. 13-17.

RECOMMENDED READING:

Anatomy of an Illness, Norman Cousins, 1983, Bantam.

5. Selye, H., "The Physiology and Pathology of exposure to Stress," Acta Inc.Medical Publications, Montreal, 1950.

6. Ringsdorf, W., et al., "Sucrose, neutrophil phagocytosis and resistance to disease," Dent. Surv., 1976, 52.
7. Alexander, M., "Oral betacarotene can increase the number of OKT4 + cells in human blood," Immunology Letters, 1985.
8. Sanchez, A., Reeser, J., Lau, H., et al., "Role of sugars in human neutrophilic phagocytosis," Am. J. Clin. Nutr., 1973.
9. Ibid.
10. Dillon, J.C., et al., "Nutritional value of the alga spirulina." Plants in Human Nutrition, 1995, vol 77. pp. 32-46.
11. Journal of the American Medical Association, Nov. 3, 1993.
12. The Lancet, Sept. 10, 1994.
13. Krause, M.V. and Mahan, K.L., Food, Nutrition, and Diet Therapy, 7th Ed., W.B. Saunders, Philadelphia1984.
14. Beisel, W., et al., "Single-nutrient effects of immunologic functions," J.A.M.A., 1981.
15. Hendler, S.S., 1990, "The Doctor's Vitamin and Mineral Encyclopedia," Simon and Schuster.
16. Infection and Immunity, Dec. 1984.

Chapter Eight

1. Meneely, G., et al., "High sodium-low potassium environment and hypertension," Am. J. Card., 1976, 38.
2. Richardson, S., Discover, July 1994.
3. Kardinaal, et al., The Lancet, Dec. 1994.
4. Panganamala, R.V., "The effects of vitamin E on arachidonic acid metabolism," Annals of the New York Academy of Sciences, 1982.
5. Murray, M., and Pizzorno, J., 1991, Encyclopedia of Natural Medicine, Prima Pub., Rocklin, Calif.
6. Ibid.
7. Gubner, R., "Vitamin K therapy in menorrhagia," South. Med. J., 1944.
8. Fabris, N., Amadio, L., Licastro, F., et al., "Thymic hormone deficiency in normal aging and Down's syndrome: is there a primary failure of the thymus?," Lancet, 1984.
* Citizens for safe drinking water, 800-728-3833.
9. Nordstrom, J.W., et al., "Trace mineral nutrition in the elderly," Am.J.Clin.Nutr., 1982.

10. Kraus, M.V., et al., Food, Nutrition and Diet Therapy, 7th Ed., W.B. Saunders, Philadelphia 1984.
11. Iseri, L., and French, J., "Magnesium: nature's physiologic calcium blocker," Am. Heart. J., 1984.
12. Brock, T.D., et al., 1984 Biology of Microorganisms, 4th Ed., Prentice-Hall, Inc.
13. The Lancet, 1994.
14. Opie, L.H., "Role of carnitine in fatty acid metabolism of normal and ischemic myocardium," Am. Heart. J., 1979.
15. Curtiss, L.K. and Plow, E.F., "Interaction of plasma lipoproteins with human platelets," Blood, 1984, 64.
16. The Lancet, 1994.
17. Weindruch, R., Scientific American, January 1996, 274(1).

Chapter Nine

1. Kirschman, J.D., et al., J. Nutrition Almanac. McGraw-Hill, 1984.
2. Gibney, M., "The effect of dietary lysine to arginine ratio on kinetics in rabbits," Athero., 1983.
3. Caruso, I., et al., "Antidepressant activity of S-adenosyl methionine," Lancet, 1984.
4. Pardridge, W., "Regulation of amino acid availability to the brain," in Wurtman, R., and Wurtman, J. (eds.), Nutrition and the Brain, vol. 1, Raven Press, New York, NY, 1979, pp. 142-204.
5. Gibson, C., "Tyrosine for depression," Adv. Biol. Psychiat., 1983, 10.
6. Ibid.
7. Sagawa, T., Ishida, H., et al. "Tryptophan 2, 3-dioxygenase-like activity of manganese," Journal of Molecular Catalysis, May 1993.
8. Seltzer, S., et al., "The effects of dietary tryptophan on chronic maxillofacial pain and experimental pain tolerance," Journal of Psychiatric Research, 1982.
9. The Lancet, May 29, 1993.
10. The Lancet, May 29, 1993.
11. Zucker, D., et al., "B_{12} deficiency and psychiatric disorders: a case report and literature review," Biol. Psychiatry, 1981.
12. Science News, Sept 2, 1995.

13. Ross, C.E. and Hayes, D., "Exercise and psychological well-being in the community," Am. J. Epidemiology, 1988.

Chapter Ten

1. Zile, M.H. and Cullum, M.E., "The function of vitamin A: current concepts," Proc. Soc. Exp. Biol. Med., 1983, 172.
2. Wright, S. and Burton, J., "Oral evening primrose oil improves eczema," Lancet, 1982.
3. Bittiner, S.B., et al., "A double-blind, randomized, placebo-controlled trial of fish oil in psoriasis," Lancet, 1988.
4. Simon, S.W., " Vitamin B_{12} therapy in allergy and chronic dermatoses," J. Allergy, 1951.
5. McCarthy, M., "High chromium yeast from acne" Med. Hypothesis, 1984.
6. Michaelsson, G., Vahlquist, A., and Juhlin, L., "Serum zinc and retinol-binding protein in acne," Br.J.Dermatol., 1977.
7. Lehninger, A. *Principles of Biochemistry,* Worth Pub. 1982.
8. Michaelsson,G., and Edqvist, L., "Erythrocyte glutathione peroxidase activity in acne vulgaris and the effect of selenium and vitamin E treatment," Acta Derm.Venerol., 1984, 64, pp. 9-14.
9. Haddox, M., et al., "Retinol inhibition of ornithine decarboxylase induction and G1 progression in CHD cells," Cancer Research, 1979.
10. Nielson, F.H., "Boron–an overlooked element of potential nutrition importance," Nutrition Today, 1988, Jan/Feb.

Chapter Eleven

1. Rubenstein, E., and Federman, D., *Scientific American Textbook of Medicine, Scientific American,* New York , NY, 1985.
2. Worthington, B.S., Meserole, L. and Syrotuck, J.A., "Effect of daily ethanol ingestion on intestinal permeability to macromolecules," Dig. Dis., 1978.
3. Majumdar, S.K., Shaw, G.K., and Thompson, A.D. "Changes in plasma amino acid patterns in chronic alcoholic patients during ethanol withdrawal syndrome: their clinical applications," Med. Hypoth., 1983.
4. Fischer, J.E., Rosen, H.M., Ebeid, A.M., et al., "The effect of normalization of plasma amino acids on hepatic encephalopathy," Surgery, 1976.
5. Ibid.

6. Rogers, L.L., et al., "Voluntary alcohol consumption by rats following administration of glutamine," J.Biol.Chem., 1955.
7. Miller, L.G., Goldstein, G., Murphy, M. and Ginns, L., "Reversible alterations in immunoregulatory T-cells in smoking," Chest, 1982.
8. Majumdar, S.K., et al., "Changes in plasma amino acid patterns in chronic alcoholic patients during ethanol withdrawal syndrome: their clinical applications," Med. Hypoth., 1983.
9. Ireland, M.A., et al., "Acute effects of moderate alcohol consumption on blood pressure and plasma catecholamines," Clin.Sci., 1984.
10. Kershbaum, A., et al., "Effect of Smoking and Nicotine on adrenocortical secretion," J.A.M.A., 1968, 203.
11. Branchey, L., Branchey, M., Shaw, S. and Lieber, C.S., "Relationship between changes in plasma amino acids and depression in alcoholic patients," Am. J. Psych.,1984, 141.
12. Ibid.
13. Baron, J., "Smoking and estrogen-related disease," A.J. Epid., 1984.
14. Dutta, S.K., et al., "Selenium and acute alcoholism," Am. J. Clin. Nutr.,1983.
15. Suematsu, T., et al., "Lipid peroxidation in alcoholic liver disease in humans," Alcoholism Clin. Exp. Res., 1981.
16. Diluzio, N.R., "A mechanism of the acute ethanol-induced fatty liver and the modification of liver injury by antioxidants," Lab.Invest., 1966.
17. Yunice, A.A. and Lindeman, R.D., "Effect of ascorbic acid and zinc sulphate toxicity and metabolism," Proc. Soc. Exp. Biol. Med.,1977.
18. Wu, C.T., Lee, J.N., Shen, W.W., and Lee, S.L., "Serum zinc, copper, and ceruloplasmin levels in male alcoholics," Biol. Psy., 1982.
19. Bode, J.C., et al., "Jejunal microflora in patients with chronic alcohol abuse," Hepato-gastro., 1984.
20. Yunice, A.A., et al., "Ethanol-ascorbate interrelationship in acute and chronic alcoholism in the guinea pig," Proc. Soc. Exp. Biol. Med., 1984.
21. Kay, R.A., "Microalgae as food supplement," Clinical Reviews in Food Source and Nutrition, Vol. 30 (1991) p. 566.
22. Davis, R.E. and Nichol, D.J., "Folic acid," Int. J. Biochem., 1988.
23. Beattie, A., et al., "Blood-Lead and hypertension," Lancet, 1976.

Chapter Twelve

1. Burch, G.E., et al., American Heart Journal, 1971.
2. Kohman, E.F. et al., "Comparative experiments with canned, home cooked and raw food diets," Journal of Nutrition 14:9-19 (1937)
3. Howell, E., "Enzyme Nutrition: The Food Enzyme Concept," Avery Pub., Inc. 1985.
4. Korenchevsky, V., et al., J. Pathology and Bacteriology, 54:13-24 (1942).
5. Howell, E., "Enzyme Nutrition: The Food Enzyme Concept," Avery Pub., Inc. 1985.

Chapter Thirteen

1. Rosenberg, E. and Belew, P., "Microbial factors in Psoriasis," Arch. Dermatol., 1982, 118, pp. 1,434- 44.
2. Petrakis, N.L. and King, E.B., "Cytological abnormalities in nipple aspirates of breast fluids from women with severe constipation," Lancet, 1981, ii, pp. 1,203-5.
3. Rosenberg, E. and Belew, P., "Microbial factors in Psoriasis," Arch. Dermatol., 1982, 118, pp. 1,434-44.
4. Morales, T.I., et al., "The effect of lipopolysaccharides on the biosynthesis and release of proteoglycans from calf articular cartilage cultures," J. Biol. Chem., 1984, 259, pp. 6,720-9.
5. Melnykowycz, J. and Johansson, K.R., "Formation of amines by intestinal microorganisms and the influence of chlortetracycline," J. Exp. Med., 1955,101, pp. 507-17.
6. Crook, W.G., *The Yeast Connection,* Professional Books, Jackson, TN, 1984.
7. Goff J.S. "Infectious causes of esophagitis." Ann Rev Med, 1988, 39, pp. 163-9.
8. Remington, D.W. and Hige, B.W., "Back to Health: A comprehensive medical and nutritional yeast control program," Vitality House Inter. 1989.
9. Tufts University Diet and Nutrition Letter, May 1992.
10. Galland, L., "Nutrition and Candidiasis," J. Orthomol. Psychiatry, 1985 15, pp. 50-60.
11. Collins, E.B. and Hardt, P., "Inhibition of Candida albicans by Lactobacillus acidophilus," J. Dairy Sci., 1980, 63, pp. 830-2.
12. Ibid.
13. Barlow, A.J.E., et al., "Factors present in serum and seminal plasma which promote germ-tube formation and mycelial growth of Candida albican." J Gen Microbiology, 1974, 82, p. 271.

14. Baccarini, M., et al., "In vitro natural cell-mediated cytotoxicity against Candida albicans: macrophage precursors as effector cells." J. Immun. April 1985, 134, pp. 2,658-65.
15. Iwata, K., "A review of the literature on drunken symptoms due to yeasts in the gastrointestinal tract." Proceedings of the Second Interational Specialized Symposium on Yeasts. Univ. Tokyo Press 1972, pp 260-68.
16. Abe, F., et al., " Experimental candidiasis in liver injury," Mycopathologia, 1987, 100, pp. 37-42.
17. Kobayashi, R.H., Rosenblatt, H.M., et al., "Candida esophagitis and laryngitis in chronic mucocutaneous candidiasis." Pediatrics, 1980, 66, pp. 380-84.
18. Svec, P., "Mechanism of action of glycoprotein from Candida albicans." J.Hyg.Epidemiol.Microbial Immun., 1974, 18, pp. 73-76.
19. Smith, R.P., "The implantation of enrichment of Bacillus acidophilus and other organisms in the intestine." Brit. Med. J. Nov. 22, 1924; pp. 950.
20. Truss, C.O., "Metabolic abnormalities in patients with chronic candidiasis; the acetaldehyde hypothesis." J. Orthomolecular Psy. 1984;13/2:24-25.
21. Miller, C., "Chemical Susceptibility and Candida albicans." Human Ecologist 1980; 3-11.
22. Young, G. et al, "Interactions of oral strains of Candida albicans and lactobacilli." Jour. Bacteriology 1956; 72;525-9.
23. Lombardi, G. et al., "Mechanism of action of an antigen nonspecific inhibitory factor produced by human T-cells." Cell Immunity 1986; 98(2):434-43.
24. Young, G., et al., "Interactions of oral strains of Candida albicans and lactobacilli." Jour. Bacteriology 1956; 72;525-9.
25. Beck, K., et al., "Yersina enterocolita infection and thyroid disorders." The Lancet 1974; 2:951-2.

Chapter Fourteen

1. Shahani, K., et al., Journal of Dairy Science 56:531-43 (1973).
2. Robbins, S.L., et al., *Pathological Basis of Disease*, Saunders,Philadelphia, Penn.,1984, pp. 847-8.
3. Messer, M., et al., "Studies of the mechanism of destruction of the toxic action of wheat gluten in coeliac disease by crude papain," *Gut,* 1964.
4. Science News, April, 1992.

5. Ransberger, K., "Enzyme treatment of immune complex diseases," Arthritus Rheuma. 1986, 8, pp.16-19.
6. Warshaw, A.L., Walker, W.A. and Isselbacher, K.J., "Protein uptake by the intestine: evidence of intact macromolecules," Gastroenterology, 1974, 66, pp. 987-92.
7. Rosenberg, I.H., et al., "Nutritional aspects of inflammatory bowel disease," ann. Rev. Nutr., 1985, 5, pp. 463-84.
8. Ward, C.G., "Influence of iron infection," Am.J.Surg.,1986, 151, pp. 291-5.
9. Henriksson, K., et al., "Gastrin, gastric acid secretion, and gastric microflora in patients with rheumatoid arthritus," Annals of the Rheumatic Diseases, 1986,45,pp. 475-83.
10. Commings, W.A. and Williams, E.W., "Transport of large breakdown products of dietary protein through the gut wall," *Gut,* 1978, 19, p.715.
11. Reinhardt, M.C., "Macromolecular absorption of food antigens in health and disease," J.Allergy, 1984, 53, p. 597.
12. Parke, A.L., "Gastrointestinal disorders and rheumatic diseases, *Current Opinion in Rheumatology*, Jan 1993; Henriksson, A.E., et al., "Small intestinal bacterial overgrowth in patients with rheumatoid arthritis," Annals of the Rheumatic Diseases, 1993 July, 52 (7): 503-10.
13. Ryle, J. and Barber, H., "Gastric analysis in acne rosacea," The Lancet, 1920, ii, pp. 1,195-6.
14. Rubenstein, E. and Federman, D.D., *Scientific American Medicine*, Sci. Amer., New York, 1988.
15. Friend, B.A., and Shahani, K.M., "Nutritional and Therapeutic aspects of Lactobacilli." Journal of Applied Nutrition 1984; 36/2:127-28.

Appendices

1. Rietschel, T. R., and Brade, H., August 1992. "Bacterial Endotoxins." Scientific American.
2. Fay, P. editor, *Cyanobacteria*, 1987. Elsevier Science Pub.
3. Nair, P., et al., "Evaluation of chemoprevention of oral cancer with Pirulina fusiformis." Nutrition and Cancer, 1995, pg. 197-202.
4. Breinholt, V., et al., "Dietary chlorophyllin is a potent inhibitor of aflatoxin B_1 hepatacarcinogenesis in rainbow trout." Cancer Research, 1995 Jan 1, 55(1): 57-62
5. Bitterman, N. et al., "Betacarotene and CNS oxygen toxicity in rats." Journal of Applied Physiology, 1994 March, 76(3):1070-6.

6. Hopewell, J.W., "The modulation of radiation-induced damage to pig skin by essential fatty acids." British Journal of Cancer, 1993 July, 68(1): 1-7.
7. Kovats, E., "Potentiation of HIV envelope glycoprotein and other immunogens by endotoxin and its molecular fragments." International Journal of Immunopharmacolgy, 1992 May, 14(4): 573-81.
8. Khalfoun, B., "DHA and EPA inhibit human lymphoproliferative responses in vitro but not the expression of T-cell surface activation markers." Scandinavian Journal of Immunology, 1996 March, 43(3): 248-56.
9. Sandyk, R., and Awerbuch, G.I., "Vitamin B_{12} and its relationship to age of onset of multiple sclerosis." International Journal of Neuroscience, 1993 July-August, 71(1-4): 93-9.
10. Herzlich, B.C. and Schiano, T.D., "Reversal of apparent AIDS dementia complex following treatment with vitamin B_{12}." Journal of Internal Medicine, 1993 June, 233 (6): 495-7.
11. Zittoun, J., "Anemias due to disorder of folate, vitamin B_{12} and transcobalamin metabolism." Revue du Praticien, 1993 June 1, 43(11): 1358-63.
12. Barbul, Adrian, et al., "Arginine stimulates lymphocyte immune response in healthy human beings." Surgery 90, 1(1981): 244-251.
13. Ducroc, R. et al., "Immunoreactive substance P and calcitonin-gene-related peptide in rat milk and in human milk and infant formulas." American Journal of Clinical Nutrition, 1995 Sept., 62(3): 554-8.
14. Hayashi, O., et al., "Enhancement of antibody production in mice by dietary Spirulina platensis." Journal of Nutritional Science and Vitaminology, 1994 Oct., 40(5): 431-41.

Mini-encyclopedia

A Glossary of Terms Used in This Book

Absorption of nutrients is the first step in good nutrition. It begins with digestion and takes place mainly through the intestinal mucosa (lining). Simple sugars and mineral ions are able to pass directly into the blood and lymph system. Larger molecules, such as carbohydrates, fats, and proteins, must be broken down into simpler sugars, fatty acids, and amino acids, respectively. This breakdown and the resultant absorption of large molecules are catalyzed by digestive enzymes and a healthy intestine that contains a high population of friendly intestinal bacteria.

Acids are molecules that, when added to water (or blood), break down into small positive hydrogen ions and larger negative ions. Acids decrease the pH of water (a pH less than 7 is considered acidic, a pH greater than 7 is considered basic, and a pH of 7 [distilled water] is considered neutral, i.e., neither acidic nor basic). Hydrochloric acid in the stomach has a pH of about 1. Fresh-water alkaline lakes (where blue-green algae flourish) have a pH between 9 and 11.

Active sites are regions on the surface of enzymes where molecules from the food we eat actually react and transform into new molecules.

Adenosine triphosphate (ATP) is a complex and profoundly important biomolecule that stores the body's chemical energy in the mitochondria of each cell.

Adrenal glands, located above both kidneys, function independently and produce hormones (messenger molecules) from their inner regions (epinephrine and norepinephrine) and outer regions (hydrocortisone).

Adrenalin. *See* epinephrine.

Albumin is an "escort" protein able to deliver nutrients to the liver by its strong electric affinity. It makes up about half of the protein found in blood serum.

ALANINE is a nonessential amino acid concentrated in muscle tissue. Alanine can be transformed into glucose when needed by muscles (or in the liver) for energy. People with hypoglycemia may have an alanine deficiency. Alanine assimilation is enhanced by the zinc and vitamin B_6 found in AFA and is known to:

1. Biostimulate the immune system by increasing the size of the thymus gland and thus lymphocyte production.
2. Be very useful for hypoglycemics by stimulating an increase in their blood sugar (glucose).
3. Function as a neurotransmitter.
4. Help build the cell walls of friendly bacteria.

ALGINATE is a food additive derived from giant kelp algae. It is used as a thickening agent in ice cream, cheese, yogurt, and whipped cream.

ALKALINE ASH is a measure of the mineral content of any food. Since AFA is 7 percent alkaline ash by weight, the mineral content of AFA algae is consequently extremely rich.

AMINO ACID METABOLISM refers to either the breakdown of individual amino acids for energy purposes or their buildup into larger proteins.

AMINO ACID OXIDASE is an enzyme used in the liver to change "old" amino acids into "new" ones. This enzyme requires riboflavin, a coenzyme, for its bioactivation.

AMINO ACID PROFILES show the proportions of amino acids in a given food compared to healthy human blood. Two foods with amino acid profiles that match most closely are the chicken egg and AFA.

AMINO ACIDS are molecules that bond together to form polypeptides or proteins. Dry AFA is about two-thirds amino acids.

AMINO SUGARS are sugar molecules that have an amino group of atoms bonded to them. These occur in the cell wall of AFA.

AMYLASE is a type of enzyme that digests starch. In the saliva it is called "salivary amylase"; in the pancreas it is called "pancreatic amylase."

ANABOLISM refers to all biochemical reactions that involve the building of larger molecules needed to construct and repair cells. The rich variety of raw materials present in AFA, such as amino acids and fatty acids, biostimulate such anabolism processes.

ANEMIA is an unhealthy condition that occurs when the amount of hemoglobin in circulating red blood cells is deficient. "Pernicious anemia" is caused by the inability to absorb vitamin B_{12} through the stomach lining. The extraordinarily high B_{12} content in AFA algae often brings relief from the debilitating symptoms of anemia.

ANIONS are ions that have a negative electrostatic charge due to the presence of extra electrons.

ANTIBIOTIC literally means "against life." Antibiotics were once considered "magic bullets" by medicine. Serious doubts about their long-term effectiveness in treating certain diseases have recently been raised because of the high amounts of antibiotics needed for treatment. Typically, antibiotics (e.g., penicillin) will dramatically diminish the population of friendly intestinal microflora and impair the function of the bowel. Birth control pills, alcohol, caffeine, some food additives, and even chlorinated water can also have similar effects. Taking supplements of friendly intestinal bacteria such as *acidophilus* can often lead to a dramatic normalization of bowel function.

ANTIBODIES are proteins formed in blood serum in response to invading viral or bacterial particles (antigens). Also called immunoglobulins, antibodies function by chemically identifying and marking invaders so that the immune system can then send out white blood cells to destroy them.

ANTIGENS are invading bacterial or viral particles that evoke the formation of antibodies (immunoglobulins) in the blood serum.

ANTIOXIDANTS are cellular biomolecules that protect organisms from the damaging effects of free radicals. AFA has an unusually high concentration and wide variety of antioxidants, such as betacarotene, superoxide dismutase, glutathione, and vitamins C and E.

APHANIZOMENON FLOS-AQUAE (AFA) is the scientists" name (in Greek, this means "invisible living flower of the water") for the blue-green algae that flourish in Upper Klamath Lake, Oregon, and serve as a nutritional protein-rich anabolic food source.

ARACHIDONIC ACID is a fatty acid found mainly in saturated animal fats. Excessive consumption should be avoided because saturated fatty acids rob the cell membrane of its healthy flexibility.

ARGININE is an amino acid that children and stress-affected people cannot biosynthesize. Dietary arginine is available from nuts, meat, cheese, and AFA. Its primary metabolic role is to stimulate the activity of a liver enzyme which initiates the urea cycle, an important biopathway that metabolizes protein. Arginine, when properly assimilated, can:

1. Biostimulate the thymus gland to enhance the immune system by producing more T-lymphocyte cells, which are effective and active in fighting infection. The activity of natural killer cells is also enhanced to inhibit tumor growth and size. Fifty years of accumulated research evidence attest to this antitumor effect.
2. Biostimulate the pituitary gland to secrete human growth hormone (GH) and produce some muscle-building and fat-burning effects.
3. Build up an "arginine reserve" to protect and detoxify the liver.
4. Increase sperm production and male fertility because human sperm is very rich in arginine.

ARTHRITIS is associated with the inflammation of joints. Poor nutrition, stress, and aging all contribute to the underproduction of natural and protective antioxidants. As a result, the synovial fluids that surround a bone in its joint accumulate excess free radicals that cause swelling and soreness. Fish oils such as eicopentaenoic acid, and AFA oils such as gamma-linolenic acid, can reduce arthritic inflammation.

ASPARAGINE is a nonessential amino acid and neurotransmitter that supplies the brain with energy. Asparagine is also an immunostimulant because it biostimulates the thymus gland.

Aspartic acid, a nonessential amino acid, has a higher concentration in the human brain than any other amino acid. It has been useful in the treatment of chronic fatigue syndrome because of its involvement in DNA/RNA synthesis. Aspartic acid has shown usefulness as a biostimulator of the thymus gland and thus as an immune system stimulant.

Aspergillus flavus; Aspergillus niger are molds found on grains, corn, and peanuts. These molds are used to produce digestive enzymes.

Atherosclerosis is a disease that causes hardening of artery walls and results in a loss of artery elasticity.

ATP (adenosine triphosphate) is an energy-storage molecule used by all cells to drive the biochemical reactions that are needed to sustain their life.

Betacarotene (provitamin A) is a free radical–quenching vitamin that is easily obtained from AFA algae, carrots, sweet potatoes, and leafy green vegetables. It is a precursor (provitamin) to vitamin A (retinol), which is essential for night vision. AFA is probably the richest betacarotene food source on Earth. Betacarotene offers powerful anticancer, antiaging, and antioxidant properties without the toxic effects of taking fat-soluble vitamin A. The betacarotene in AFA most definitely boosts the immune system by protecting the cell lining of the digestive and respiratory tracts. Apparently, the number and strength of immune cells will dramatically increase so that a variety of infections can be fought off. In general, malignant cell growth appears suppressed by betacarotene. *See* carotenes.

Bifidobacterium bifidum is (or should be) the friendly bacteria living mainly in the colon (large intestine). The colons of breast-fed children are inhabited mainly (99 percent) by this bacteria. As aging diminishes the numbers of these bacteria, so too does a person's general health decline. These friendly bacteria produce special acids which prevent the growth of some viruses and bacteria. These friendly microflora even help to prevent nitrite toxicity (present in hot-dogs and other packaged meats). It has even been effective against the bacteria that causes toxic shock syndrome.

Bile is a yellow-green bitter fluid continuously secreted by the liver, stored in the gall bladder, and used in the small intestine (duodenum) – along with pancreatic juices – to digest fats. When fats enter the small intestine, a hormone is secreted to signal this process. Bile is actually a mixture of water, pigments from hemoglobin breakdown, bile salts, and cholesterol.

Bioflavonoids are found in both AFA algae and in the pulp of citrus fruits. Once called Vitamin P, bioflavonoids are now considered important phytochemicals that can help boost the immune system. The small amount of bioflavanoids in AFA can also help to increase capillary strength so that cellular nutrients and wastes can be more efficiently transported.

Biosynthesis. *See* Synthesis.

Biotin is a water-soluble B vitamin coenzyme that is obtained from a good diet (e.g., brown rice, mushrooms, and AFA) and friendly intestinal bacteria. It is used in a variety of important enzyme systems to construct fatty acids by attaching acid functional groups to larger molecules. Because of this, biotin can produce healthy-looking hair. Biotin also improves athletic performance by helping to metabolize branched-chain amino acids.

BLOOD is a life-sustaining fluid with a majority of red blood cells (erythrocytes) and a minority of white blood cells (leukocytes) and platelets. Oxygen is carried by the red blood cells, and larger molecules (e.g., hormones, glucose, lipids) are transported in the blood plasma. The pH of healthy blood is about 7.4.

BLOOD-BRAIN BARRIER refers to the inability of most molecules other than glucose to cross the barrier from the brain's blood vessels into the brain's tissues.

BLOOMS are profuse growths of microalgae that discolor the water in which they live for a short duration. An AFA bloom occurs four times per year in Upper Klamath Lake (Oregon).

BLUE-GREEN ALGAE. *See* cyanobacteria.

BORON is a "bone-strengthening" trace element found in soil. Its concentration in soil varies depending on farming practices. Boron is obtained from apples, pears, leafy greens, and AFA and it improves calcium retention. US government-sponsored research shows that boron deficiency in humans may be common, especially because the elderly (and alcoholics) have increasing difficulty absorbing boron from their intestines. It is possible that osteoporosis, arthritis, and hypertension may be lessened with boron-rich foods such as AFA. Some researchers even claim that boron may enhance mental clarity. Artery plaque due to calcium deposits (and thus high blood pressure) seems to be lessened in the presence of boron. Citizens of countries with soil containing high quantities of boron (e.g., in Israel) have a lower incidence of arthritis than those living in countries with less rich soil (e.g., in Jamaica).

BOWEL TOXEMIA occurs when the elimination function of the colon is reduced by and impacted with the toxins of unfriendly bacteria, viruses, and parasites. Myriad health problems develop, along with swelling and inflammation of the bowel, which prevent or reduce the absorption of nutrients. Eventually, toxins enter the bloodstream and cause more problems with the circulatory and lymphatic system, even the lungs and kidneys. One way to help insure a properly functioning intestine is to use probiotics regularly. *See* leaky-gut syndrome.

BRANCHED-CHAIN AMINO ACIDS (BCAAs) refer to the essential amino acids valine, leucine, and isoleucine. About one-third of our muscle mass is made of BCCAs, so having a generous and continual supply of them can help prevent muscle loss and even build muscle mass.

BROMELAIN is a digestive enzyme found in pineapple that helps to digest vegetable protein in the gastrointestinal tract.

BROWN ALGAE, commonly known as "seaweed" or kelp, are found floating freely in ocean water.

CALCIUM is the "anti-osteoporosis" mineral element that is found in dairy products, leafy green vegetables, tofu, and especially in AFA. Swiss cheese contains about 10.0 mg per gram,

AFA contains 14.0 mg per gram. Because calcium absorption in the intestines diminishes with age, calcium will be slowly lost and cause a degenerative "porous-bone" disorder called osteoporosis. Although as much as 75 percent of dietary calcium does not get absorbed, calcium absorption is enhanced by the betacarotene and chelating amino acids found in AFA. Because some enzyme systems involving neurotransmitters are activated by calcium, the calcium in AFA has a nerve-calming effect and acts nicely as a bedtime tranquilizer. Muscle cramping is diminished. Calcium also protects against cardiovascular disease by lowering cholesterol and blood pressure.

CANDIDA ALBICANS is a common yeast present in the gastrointestinal tract of almost everyone. It can cause a variety of allergic reactions if the intestinal tract is overrun. As *Candida* continues to proliferate, the immune system weakens and problems such as irritable bowel syndrome, premenstrual stress syndrome, food sensitivities, depression, prostate pain, and even chronic carbohydrate cravings begin to develop.

CARBOHYDRATE METABOLISM begins with glucose (blood sugar) being carried to the liver. With the transformative effects of enzymes and the "energy molecule" ATP (adenosine triphosphate), glucose is phosphorylated (i.e., phosphorous and oxygen are added) and thus transformed step-wise into glycogen, a complex carbohydrate that acts as an energy reserve for later combustion.

CARBOHYDRATE DIGESTION depends upon the action of several enzymes. Salivary amylase begins the process whereby complex carbohydrates are broken into polysaccharide fragments. Pancreatic amylase (and intestinal enzymes) further breaks the fragments into individual monosaccharides.

CARBOHYDRATES may be classified as sugars, starches, or celluloses depending on the number and bonding arrangement of their constituent sugar groups. A somewhat straight chain of about 1000 sugar groups is a starch, whereas a branched-chain of similar length describes glycogen. Such long chains are also called polysaccharides. Shorter chains of two sugar groups are called disaccharides (e.g., lactose, sucrose), and individual sugar groups that are independent of each other are termed monosaccharides (e.g., glucose and fructose).

CARCINOGENS are chemical substances that initiate tumor formation.

CAROTENES or carotenoids are reddish-yellow plant pigments found abundantly in AFA algae, carrots, and leafy green vegetables. When carotene molecules are digested (oxidized and "broken" in half), they are converted from vitamin A precursors to retinol molecules, the liver-storage form of vitamin A. The most active and beneficial type of carotene is the dark-red betacarotene. Alphacarotene is closely related to betacarotene, as are the less well known carotenes, lutein and zeaxanthin.

CATALASE is an enzyme that can absorb and render harmless the dangerous peroxide free radicals created during normal metabolism.

CATALYSTS are substances that speed up a chemical reaction without being consumed. Catalysts made of protein material are called enzymes.

CATECHOLAMINES are an important family of neuropeptides that include dopamine and adrenaline. The tyrosine in AFA acts as their precursor.

CELLS are fundamental units of living matter capable of reproduction. Eukaryotic cells have a well-defined central nucleus, whereas prokaryotic cells do not.

CELL MEMBRANES are semifluid lipid and protein coverings that surround the body or cytoplasm of the cell. Billions of years ago, in the ancient oceans of Earth, such lipids gradually formed into primitive cell membranes that were necessary for the maintenance of life and the eventual formation of blue-green algae (cyanobacteria). The fatty acids that form the lipid layers of the cell membrane ultimately affect the membrane's fluidity, depending upon the actual shape and length of each fatty acid. The cell membrane of any cell needs to be "fluid and flexible" if it is to remain healthy. Cholesterol is also important to the cell membrane in that its molecules are imbedded between the fatty acid molecules of the lipid bilayer, helping to hold them together and thus impart integrity and strength to the cell membranes. There is one drawback, however: Too much cholesterol holds the membrane together too rigidly so that the membrane loses its fluidity and thereby its health. The cell membrane of AFA algae is very flexible, partly because it does not contain any cholesterol whatsoever.

CELLULOSE is a polysaccharide composed of about 1000 glucose units that are bonded together such that animal enzymes cannot break (digest) them. Cows are able to digest cellulose only because of the unique enzymes produced by the friendly bacteria in their intestines.

CEPHALINS are a class of phospholipids found mostly in the cell membranes of brain and nerve cells. Because these cells are not replaceable once they wear out, it is especially critical that such cephalins be constructed from the kind of polyunsaturated fatty acids found in raw fish, nuts, seeds, and AFA.

CHELATED MINERALS are found in blue-green algae such that the metal atoms of each mineral element (e.g., iron, zinc, magnesium) are naturally bonded to nonmetal atoms inside of larger peptides, protein, or enzyme molecules.

CHELATES are molecules – usually peptides – that can chemically bond to minerals by holding them with their nitrogen or oxygen atoms. The word chelate comes from a Greek word meaning "claw." Iron is chelated within the hemoglobin molecule by being attached to several negatively charged nitrogen atoms.

CHELATION THERAPY is a biochemical process of ridding the body of toxic metals that have been accumulated from a toxic environment. The presence of such metals (e.g., lead, nickel, mercury, cadmium) lowers resistance to disease, causes enzyme dysfunction, and imperils basic physiological functions. Most physicians use EDTA (ethylenediaminetetraacetic acid) in their chelation therapy. Although used by the U.S. Navy to help treat lead poisoning, EDTA is a synthetic drug without full FDA approval. Sulfur-containing amino acids (cysteine and methionine) found in AFA algae, along with chlorophyll and glutathione (a chelating peptide), naturally and slowly begin to reduce the amount of toxic metals in the body. This may improve memory and concentration and help to reduce arthritic and muscular pain.

CHLORELLA is a green algae used as a food only after its indigestible cell wall is removed.

CHLOROPHYLL is a photosynthetic pigment found abundantly in plants and algae, especially AFA algae. AFA contains 2 to 3 percent of this wondrous green molecule, placing our blue-green algae among the highest in chlorophyll content on Earth. It is considered by some to be a cell regenerator partially because its central magnesium atom plays an important role in so many (325) different enzyme systems. The chlorophyll in AFA may help to inhibit the mutation-promoting effects of some environmental carcinogens. It may do this by acting as an antioxidant and thus protect our DNA during cell division, a very vulnerable time in the life of any cell.

CHLORINE is a poisonous gas, but is found in nature as the chloride ion within sodium chloride (table salt). Chloride is part of our stomach acid in the form of hydrochloric acid and helps our liver to detoxify and maintain the body's acid-base balance.

CHOLESTEROL is a polycyclic molecule found in all animal fats and oils. It can chemically combine with certain proteins to form lipoproteins such as high-density lipoproteins (HDLs) – the "good" cholesterol – or low-density lipoproteins (LDLs) – the "bad" cholesterol. Cholesterol is a precursor molecule of vitamin D and a variety of steroid hormones. However, too much cholesterol can build up and can cause artery blockage.

CHOLINE is a medium-sized molecule, part alcohol and part amine, which occurs as a component of biological tissue in lecithin and acetycholine. Choline is present in all foods, so deficiencies of choline do not exist.

CHROMIUM is a trace micromineral with only about 5 milligrams distributed throughout the entire human body. As a cofactor, chromium bioactivates a variety of enzymes which initiates insulin action, controls glucose levels, and stimulates fatty acid synthesis in the liver. In addition, chromium is the "sugar-regulating" trace mineral found in brewer's yeast, black pepper and AFA algae. Refined sugars should be avoided because it causes the body to excrete its valuable chromium.

This micronutrient metal is an essential part of a relatively large glucose-regulating tripeptide molecule called the "glucose tolerance factor" (GTF). The exact structure of this molecule is presently unknown. Our mineral-poor topsoil, along with our over-refined foods, has led to an alarming depletion of chromium in the American diet. As a result, diabetes and atherosclerosis are much higher in America than in Asia, Africa, or the Middle East. In fact, diabetes has increased 600 percent since the 1940s with at least 300,000 annual related deaths in the 1990s. To make matters worse, chromium is relatively difficult to absorb. Only about 1 percent of ingested chromium even winds up in the bloodstream, the rest being excreted. Absorption may be raised to much higher levels if the metal ion is chelated, as it is in AFA. There is some evidence that refined white flour should be avoided because it removes chromium from the blood.

Chromium probably reduces adult-onset diabetes and hypoglycemic symptoms. As part of GTF, chromium helps the insulin molecule to bond to receptor sites on glucose-hungry cells, allowing glucose to efficiently migrate from the blood into the cell. This is desirable because excessive glucose circulating in the blood will often damage artery walls by oxidation.

CHRONIC describes a disease that lasts a relatively long time. Acute pains go away quickly; chronic complaints might last a lifetime.

CIRCULATING IMMUNE COMPLEXES (CICs) are incompletely digested proteins in the blood that are marked by antibodies as if they were toxic antigens. This puts a strain on the immune system. Protein-digesting enzymes from *aspergillus* are effective in breaking them up.

CIRRHOSIS is a digestive disease of the liver caused by poor nutrition and/or alcohol or drug abuse. The liver gets "scarred" internally so that its normal function (there are hundreds!) is greatly impaired.

CIS AND TRANS refer to the two different shapes that an unsaturated fatty acid can have. Each has remarkably different biochemical effects on the fluidity of the cell membrane. A "cis" type fatty acid, for example, has a twisted carbon chain and tends to stack up randomly and form flexible cell membranes. A "trans" type fatty acid has a much straighter carbon chain that tends to stack up neatly and form solids or gooey liquids that can block up blood vessels.

COBALAMINE (VITAMIN B12) is the "rejuvenator and energizer" vitamin. It can be obtained from fish, meat, liver, dairy products, and AFA algae (the richest known source). Cobalamine is used in a wide variety of enzyme systems. As a result, it can be used to synthesize hemoglobin and thus keep blood oxygenated, to increase energy and promote growth, and to maintain and repair the nervous system. Vegetarians (especially children) need to watch out for this deficiency because vitamin B_{12} is not present in plants. AFA is one way the vegetarian can prevent this detrimental deficiency. Cyanocobalamin is one of its more active forms and plays a role in red blood cell synthesis.

COBALT, an important cofactor of vitamin B_{12}, helps to biosynthesize red blood cells and repair nerve tissue. Cobalt is found in AFA, oysters, ocean fish, red meat, and liver. The high concentration of cobalt in AFA is good news for vegetarians because the U.S. soil content of cobalt is normally very low.

COD LIVER OIL is a very important fish oil because of its vitamin A and vitamin D contents. Its value, in part, probably derives from the fact that cod fish eat large quantities of algae.

COENZYMES are nonprotein molecules (usually B vitamins) that bond to protein molecules and impart to them the ability to function properly as enzymes.

COENZYME Q10, OR UBIQUINONE, removes electrons from foods (proteins, carbohydrates, and fats) and carries them directly to oxygen molecules, turning them into water. Ubiquinone is the coenzyme that makes the process of food oxidation possible. It occurs mainly in the mitochondria "energy factory" within cells.

COFACTORS are inorganic metals that are required for enzyme activity.

COLITIS is a disease in which the colon becomes irritated and inflamed. The ingestion of probiotics has been shown to have some success in treating this painful disease.

COLLAGEN is a protein that makes up most of the connective tissue between skin cells and between and inside bones.

CONJUGATED PROTEINS are proteins that can be chemically connected to fats (lipoproteins), carbohydrates (glycoproteins), or phosphates (phosphoproteins).

COPPER is an essential trace mineral found in liver, shellfish, nuts, and AFA algae. As a micronutrient, it is important along with iron in many enzyme systems as a catalyst for the biosynthesis of hemoglobin. Copper is part of the antioxidant enzyme superoxide dismutase (SOD), which may have an antiarthritic effect in the synovial joints.

CYANOBACTERIA are blue-green prokaryotic algae such as *aphanizomenon flos-aquae* that carry out photosynthesis. They are the simplest of the algae in that they do not have a nuclear membrane. Although there are probably about 1000 different species, only AFA, *spirulina*, and a few other species are edible.

CYANOPHYCIN STORAGE GRANULES are found inside the AFA cell and are composed of a polymer of aspartic acid and arginine, two amino acids that stimulate the immune system.

CYSTINE is an important sulfur-containing nonessential amino acid that is synthesized from cysteine in the liver and then used in a variety of interesting and powerful ways. The shapes of large and complex protein molecules depend on the placement of cystine. By use of its sulfur atoms it can hold the otherwise straight chain of amino acids in more complex patterns. Cystine is sometimes considered an "antiaging" nutrient because of its detoxifying properties. Cystine helps to detoxify carcinogens by using its sulfur atoms to absorb the damaging effects of free radicals.

CYTOPLASM refers to the cellular contents between the cell membrane and the nucleus.

DERMATITIS is a skin disorder caused by irritants in the environment and aggravated by stress and fatigue. Eczema, with its flaky and itchy blisters, is caused in part by a deficiency of B vitamins and/or amino acids. AFA algae, being rich in both, is a good preventative for dermatitis.

DIGESTION involves the use of the alimentary canal (mouth, stomach, intestines) and the gastrointestinal (GI) tract, supported by digestive organs (salivary and intestinal glands, pancreas, liver, gall bladder) where enzymes and hydrochloric acid break down and convert (1) fats to fatty acids, (2) proteins to amino acids, and (3) carbohydrates to sugars.

DIGESTIVE ENZYMES are used to break down food. They are either endogenous or exogenous, depending on their origin. When food is cooked, inherent and exogenous digestive enzymes are destroyed (denatured). This forces our digestive system to compensate by producing more endogenous digestive enzymes.

DISACCHARIDES are relatively simple carbohydrates that break down into two simpler sugar molecules. Ordinary table sugar is a disaccharide called sucrose and catabolizes into glucose (blood sugar) and fructose (fruit sugar).

DNA (DEOXYRIBONUCLEIC ACID) is found in the central region of all cells. DNA biochemically transfers genetic information to successive cellular generations by making a copy of itself (replicating). The DNA molecule itself is a "double helix" of molecular groups (sugar, phosphate, and nitrogen bases) that are arranged into chains from which the proteins (and enzymes) needed for the life of a cell may be constructed. These chains, sometimes called genes, are also grouped together into larger units called chromosomes. The DNA in AFA algae consists of a single circular molecule.

DOPAMINE is an important neurotransmitter that helps to promote memory and improve alertness. It is derived from the amino acid tyrosine.

E. COLI (ESCHERICHIA COLI) BACTERIA are rod-shaped bacteria that live in human intestines. They are considered "friendly" in that they provide us with vitamin K and some B vitamins. They could, however, be considered "unfriendly" at times because they can produce infection if introduced into open wounds.

EICOSAPENTAENOIC ACID (EPA) is a semiessential fatty acid present in cold-water fish such as cod, mackerel, salmon, and AFA. Greenland Eskimos have greatly reduced cardiovascular disease because they eat such EPA-containing fish, which are known to dine on large quantities of blue-green algae.

ELECTROLYTES are substances that, when added to water, break up or ionize into positively and negatively charged electrical particles called ions.

ENDOCRINE GLANDS are among those six ductless glands (e.g., thyroid, adrenals, pancreas) that secrete hormones carried throughout the bloodstream. These hormones are incredibly powerful even though their amounts are measured in micrograms.

ENDORPHINS are morphine-like neuropeptides. They may be formed in small quantities following the digestion of low molecular weight peptides found in AFA.

ENDOTOXINS are powerful lipopolysaccharide (LPS) substances found within the cell walls of some bacteria and blue-green algae (cyanobacteria). LPS endotoxins often biostimulate the immune system by increasing the disease-fighting activity of white blood cells.

ENZYMES are biochemical protein catalysts that speed up reactions in the body without themselves being consumed or chemically altered. Most enzymes also require a vitamin coenzyme and/or a metal ion cofactor to be bioactive and functional. Each cell requires several thousand different enzymes. Typically, enzymes are named by adding "ase" to the name of the substrate molecule. For example, maltase catalyzes the hydrolysis (breakdown) of maltose, lactase hydrolyzes lactose, and lipase digests lipids.

EPINEPHRINE (OR ADRENALIN) is an enzyme activator secreted from the adrenal glands (one of six endocrine glands) that causes glucose to be quickly released from the liver into the bloodstream, allowing for an emergency "fight or flight" situation. Each adrenalin molecule is responsible for the release of roughly 30,000 glucose molecules. The essential amino acid tyrosine (with vitamin C) is needed to biosynthesize adrenalin.

ERYTHROCYTES. *See* red blood cells.

ESSENTIAL refers to those molecules that are necessary for life processes but impossible for the body to produce. As a result, essential molecules (e.g., essential amino acids, essential fatty acids) must be supplied by the diet.

ESSENTIAL AMINO ACIDS are amino acids that humans cannot biosynthesize at the rate needed. There are eight of them from among the twenty amino acids needed for protein synthesis. The quality of any protein source is determined by the amount and proportion ("amino acid profile") of these eight essential amino acids.

ESSENTIAL FATTY ACIDS (EFAs) are fatty acid molecules which cannot be biosynthesized by humans (hence the word "essential"). They must be procured from food. Linoleic and linolenic fatty acids – once called vitamin F – are essential to survival. EFAs provide energy, insulation, and structure while maintaining fluid and flexible cellular membranes. They are precursors of prostaglandins, which are required for hormone-like purposes. The World Health Organization recommends that 3 to 5 percent of the diet include these essential fatty acids. AFA algae is rich in such compounds.

EUKARYOTIC cells have a true nucleus and a variety of organelles, all of which are surrounded by membranes. Eukaryotic cells evolved from the more basic prokaryotic cells, exemplified by cyanobacteria, the blue-green algae.

FATS are made up of glycerol (a sweet polyalcohol) bonded to one, two, or three fatty acids. The resulting fats – called oils in the liquid form – are digested mostly in the small intestine, where they are broken down by pancreatic and intestinal enzymes. After being delivered by the blood to muscle and liver tissue, fats are either oxidized to supply 70 percent of the body's energy needs or stored in fat tissue for later use.

FATTY ACID ENZYMES are used in the biosynthesis of nonessential fatty acids. A family of seven such enzymes – collectively called "fatty acid synthetase" – is used.

FATTY ACIDS typically have a long hydrocarbon chain that always ends with an organic acid group of atoms. The molecule may be monounsaturated or polyunsaturated. Linoleic and linolenic acids are both abundant polyunsaturated fatty acids. They are called "essential" when they cannot be biosynthesized by humans.

FIBER refers to indigestible toxin-absorbing polysaccharides, such as cellulose, hemicellulose, and pectins, found within cell walls. AFA contains about 5 percent dietary fiber, ranking it high in those polysaccharides needed by the colon to eliminate toxic wastes.

FIXATION refers to the chemical incorporation of carbon dioxide and/or nitrogen from the air into cellular material by plants or algae.

Fluorine, like chlorine, is a poisonous gas. However, as the negatively charged fluoride *ion* it is known to be potent against tooth decay and possibly to help slow down osteoporosis. Fluoride ions are not yet considered essential to health. They are found naturally in most food, water, and soil. Apparently, fluoride in our water (or our algae) is protective against osteoporosis because of the biostimulation of new bone growth and the concurrent hardening of the bone itself.

Folic acid is the "DNA synthesizer" vitamin molecule that is obtained from a good diet of AFA algae and fresh leafy green foliage. Deficiency of this vitamin may lead to anemia, weakness, and irritability. Alcoholics and elderly people are at risk because of intestinal malabsorption problems. Because of this, the ingestion of a combination of AFA, plant enzymes, and probiotics is recommended. Folic acid also appears to improve mental function (possible IQ increase) in children.

Food additives range from coloring agents to antimicrobials, and number around 2000. Red Dye Number 3 is probably neurotoxic because it may interfere with neuropeptide function; Blue Dyes 1 and 2 may be carcinogenic. Some people in state and city governments are taking notice of the potential harm of these food additives.

Food-enzyme stomach region refers to the upper portion of the stomach where predigestion occurs for about one hour. Predigestion is followed by gastric acid secretions, which then deactivate the endogenous digestive enzymes at work.

Free radicals are molecules (or ions) that are highly reactive chemical cousins of molecular oxygen and are often generated by otherwise harmless biochemical processes. They are disruptive to all living organisms because they do not have all of their electrons paired up as do other more stable molecules. Thus, free radicals will steal back their missing electrons from such important biomolecules as DNA or membrane lipids, causing disease and speeding up the aging process. Many antioxidant substances in AFA algae are able to absorb and disarm such free radicals.

Friendly bacteria are those symbiotic microorganisms that live in the gastrointestinal tract of humans (and animals) and contribute to good health. The 400 or so species present in humans weigh roughly 4 pounds and number in the trillions, equaling or exceeding the total number of cells in the body. Typically, one-third of dry fecal matter is actually dead bacteria, friendly or otherwise. Antibiotics such as penicillin and steroids (e.g., prednisone, cortisone, and birth control pills) greatly reduce the population of these important microbacterial flora.

Germanium is an important micromineral with a variety of positive healthful benefits. Sometimes it is useful in controlling the virus in chronic Epstein-Barr disease. Organic germanium seems to stimulate the immune system and have anticancer activity.

Gamma-linolenic acid (GLA) is a semiessential fatty acid that is critical to the maintenance of a healthy and fluid cell membrane. GLA is enzymatically biosynthesized from linoleic acid – an essential fatty acid – and is present to varying extents in evening primrose oil, borage oil, mother's milk, and most abundantly in AFA algae. GLA effectively lowers high cholesterol levels over 100 times more efficiently than does linoleic acid.

Glucagon is a hormone that increases the blood sugar (glucose) level by stimulating the liver to break down its polysaccharide glycogen store into glucose.

Glucose is the blood sugar and monosaccharide unit used in building glycogen, the "animal starch" stored in the liver as a source of energy. Glucose supplies energy to the brain because it so easily crosses the "blood-brain barrier."

Glucose tolerance factor (GTF) refers to a glucose-regulating molecule that is bioactivated by chromium. GTF helps to transfer glucose from the blood to individual cells. The chromium, niacin, and amino acids in AFA may be useful for the construction of GTF.

Glutamic acid is a nonessential amino acid and "brain fuel" readily obtainable from animal, vegetable, or algae protein. It is concentrated in the human brain's memory center (hippocampus) as a "stimulant" neurotransmitter. It is for this reason that glutamic acid can improve mental functioning. Glutamic acid also helps to reduce alcohol and sugar cravings.

Glutamine is a nonessential amino acid derived from glutamic acid. It plays a major role in DNA synthesis. Glutamine, like glutamic acid, is a neurotransmitter "brain fuel." Its concentration in blood is typically much higher than any other amino acid.

Glutathione is a tripeptide coenzyme that is made from glutamic acid, cysteine, and glycine – all amino acids present in AFA algae. It can function as a "biological electrical conductor" by transferring electrons from one molecule to another. Glutathione is also considered an antioxidant and is able to protect the body from attack by dangerous free radicals.

Glycerol or glycerin is a sweet, syrupy, polyalcohol liquid.

Glycine is a nonessential amino acid that is important in brain chemistry and the synthesis of DNA, hemoglobin, and collagen. Found in protein foods, it is the simplest amino acid, tastes like glucose (hence its name), and has a calming effect. Glycine receptor sites are found throughout the central nervous system. Glycine has a variety of nutritional uses:

1. Can lower cholesterol and triglyceride fats in the blood.
2. Detoxifies benzoic acid, a common food additive.
3. Stimulates the production of glutathione, probably the most important endogenous antioxidant in all living organisms.
4. Stimulates the secretion of glycogen, and therefore balances blood sugar levels.
5. Heals wounds because skin collagen requires high amounts of glycine.

Glycogen is the "animal starch" stored in the liver. It serves as our carbohydrate source of energy. Glycogen can be broken down into glucose or energy by a variety of life-sustaining enzymes.

Glycogen granules are naturally found inside the AFA cell and are composed of polymers of glucose that serve as a carbohydrate energy reserve.

GLYCOLIPIDS are fatty acids that are chemically bonded to sugar molecules and found mainly in brain tissues. AFA contains these important brain chemicals.

GLYCOPROTEINS consist of amino acids and sugar molecules chemically bonded together. They are important in the structure of skin, blood vessels, and cell membranes. Glycoproteins "actively transport" metals in and out of cells on the surface of the cell membrane.

GREEN ALGAE are mostly fresh-water inhabitants that contain a high concentration of chlorophyll. *Chlorella* algae are inedible unless their cellulose cell walls are removed.

HEALTHY COLON MUCOSA is achieved when the cell surface lining (mucosa) efficiently secretes necessary hormones and lubricants, aggressively prevents toxin absorption, and systematically allows usable nutrients into the bloodstream. Healthy intestinal peristalsis that produces about one contracting movement every four seconds is imperative for regular bowel movement and a properly functioning colon. A regular regimen of probiotics is strongly recommended. Some of the many symptoms that can be corrected with probiotics are:

- Backache and headache
- Fatigue and concentration loss
- Constipation, gas, bloating and indigestion
- Bad breath and coated tongue

HEAVY METALS such as lead, mercury, cobalt, and cadmium are environmental toxins that cause a variety of physical, emotional, and even mental problems. The powerful short chain peptides in AFA may help to gradually remove these contaminants by the process of chelation.

HEMOGLOBIN consists of a complex protein molecule (globin) bonded to a nonprotein portion (heme) which contains iron that is capable of bonding to oxygen long enough to deliver it to the animal cells that require it. This ability facilitates the process of delivering oxygen to and removing carbon dioxide from each cell. There are 574 amino acid molecules required to biosynthesize hemoglobin. The heme ring is nearly identical to chlorophyll, except iron is replaced by magnesium.

HETEROCYSTS are specialized blue-green algae cells within which nitrogen fixation takes place so that amino acids may be enzymatically constructed.

HISTIDINE is a conditionally essential amino acid needed by children and stress-affected people. Dietary histidine is available from most animal and plant protein, but not very much is really known about it other than its role in boosting the immune system and in producing red and white blood cells. Histidine is also a mineral chelator and thus enhances intestinal absorption of minerals – especially zinc and copper.

HOMEOSTASIS refers to that natural and healthy tendency associated with the internal chemistry of the body to continually maintain a balanced, resilient, dynamic, and fluid internal stability. This state occurs when there is a coordinated response of specific parts of the body – organs, tissues, hormones, enzymes – to intelligently respond to the stresses of outside stimuli so that normal function is not disturbed or "dis-eased."

HORMONES are regulatory molecules secreted by specialized cells, such as those of the endocrine glands, throughout the body. Hormones dramatically affect such bodily functions as digestion, growth, and sexual development.

HYDROCHLORIC ACID (HCl) is a digestive acid in the stomach. It is secreted by the gastric cells of the stomach wall and converted to pepsin, a protein-digesting enzyme. Since HCl is strongly acidic (pH = 1), invading organisms can be destroyed. The HCl also helps enzymes to break the bonds within carbohydrates and proteins.

HYDROLYSIS refers to the chemical process by which chemical bonds between small molecular units (e.g., amino acids, sugars) in a polymer are enzymatically broken by the action of water.

HYPERVITAMINOSIS, OR VITAMIN TOXICITY, is a disorder brought on by ingesting too many vitamin supplements containing excessive fat-soluble vitamins such as A and E.

IMMUNOSTIMULATION, or IMMUNE SYSTEM ENHANCEMENT, occurs when there is an increase in white blood cell activity. Healthy probiotics with a strong population of friendly *acidophilus* bacteria is a prerequisite. Pathogenic (harmful) bacteria such as *Candida* and *Clostridium*, which can impair the immune system, are destroyed by the lactic and acetic acids produced by such friendly bacteria.

INSULIN is a protein secreted by the pancreas (endocrine gland) in response to the rise in blood sugar (glucose) that follows carbohydrate assimilation. Insulin is the only hormone that initiates the conversion, in the liver, of monosaccharide glucose to polysaccharide glycogen.

INTERFERON is a protein, produced by white blood cells, that provides protection from invading viruses.

IODINE is a trace mineral needed by the thyroid (endocrine) gland to biosynthesize a variety of hormones which control energy expenditure and how fast betacarotene is converted to vitamin A. The best source of dietary iodine is kelp and ocean fish. AFA provides a small daily amount. Unfortunately, fast food restaurants use too much "iodized" salt in their food. Such high doses of iodine often cause acne.

IRON is the "rosy cheek" trace mineral absolutely essential for human survival. Fatigue, irritability, depression, and low enthusiasm are all typical symptoms of iron deficiency. Iron has been used to bioactivate enzymes, which have been essential in DNA synthesis by ancient microorganisms like blue-green algae for 3–4 billion years. Iron also biostimulates the immune system by generating free radical "bullets" that can help to destroy invading bacteria.

Iron is available in red meat, salmon, brewer's yeast, green vegetables, and AFA. Iron absorption takes place via the intestinal walls over a period of roughly three hours. Vitamin C increases iron absorption, as does a small amount of copper, cobalt, or manganese, all of which are present in AFA. Hemoglobin molecules give blood its red color and contain iron ions that are used to transport oxygen to every cell of our body.

ISOLEUCINE is an essential branched-chain amino acid (BCAA) that is available in meat, fish, and AFA. Concentrated mostly in our muscle tissue, it is biochemically used in energy production and muscle building by those who have been ill for a long time and need to restore lost muscle mass. It seems to reduce muscle twitching and tremors by stabilizing blood sugar. Isoleucine is also useful in treating liver damage from alcoholism and neurological side-effects from an impaired liver.

JAUNDICE refers to the yellow tint imparted to the skin and eyes by an excess of yellow bile pigment (bilirubin), the natural breakdown product of red hemoglobin. This can occur when red blood cells are dying faster than the liver can process them.

KREBS CYCLE, or citric acid cycle, is a series of important biochemical events wherein specific enzymes oxidize proteins, carbohydrates, and fatty acids for energy production. The cycle may be thought of as a "wheel" that turns faster (and thereby produces more energy) as the amount of enzyme is increased. The energy produced in this cycle supplies us with 90 percent of our needs.

LACTASE is a digestive enzyme secreted by gastrointestinal mucous cells that "breaks" (hydrolyzes) lactose (milk sugar) into glucose and galactose. Some people have permanently lost this enzyme and so have a condition called "lactose intolerance."

LACTIC ACID is an acid produced by the bacterial fermentation of lactose (milk sugar) and during muscle contraction.

LACTOBACILLUS ACIDOPHILUS is a type of friendly bacteria that primarily inhabits the small intestine in humans and animals and manufactures the enzyme lactase, which digests sugars such as lactose (milk sugar). *Acidophilus* bacteria also produce lactic acid and natural antibiotics, which kills off *Candida* yeast as well as the invading population of 27 different bacteria, including *salmonella*. They even help lower cholesterol. Unfortunately, the use of certain medicines and drugs can drastically reduce the population of these friendly bacteria in the intestine.

LACTOBACILLUS BULGARICUS may be used to make yogurt, along with *Streptococcus thermophilus*, another friendly intestinal bacteria. Both types produce antibiotics and lactic acid, which kill harmful bacteria, thus creating a better intestinal environment for the proliferation of other friendly bacteria, such as *Lactobacillus acidophilus* and *Bifidobacterium bifidum*.

LACTOSE is milk sugar, a disaccharide of glucose and galactose.

LACTOSE INTOLERANCE is the inability to digest lactose due to the absence of the enzyme lactase. Because lactase diminishes with age, "lactose intolerance" can develop in many adults and lead to stomach cramps, nausea, and even diarrhea. Most non-Caucasian adults have this condition and should avoid or minimize milk product ingestion.

LAKE CHAD is an alkaline African lake in Chad, Africa, where the local Kanembu tribal people gather blue-green algae and dry it into protein-rich cakes called "dihe."

LEAKY-GUT SYNDROME occurs when toxins, bacteria, and/or undigested food molecules pass into the bloodstream due to complications from bowel toxemia. Such molecules cause an immune reaction when antibodies are formed against them. Unfortunately, healthy tissue is also attacked. This leads to degenerative diseases such as rheumatoid arthritis and thyroid disease. Since these toxins can damage and deform the DNA structure in healthy cells, cancer may develop as well.

LECITHIN is a phospholipid made up of a glycerol molecule chemically bonded to two fatty acids and one phosphate group. As a result, lecithin acts as an emulsifier and may help to dissolve cholesterol deposits. The lecithin in AFA helps to create a "healthy cell membrane." Letithin deficiency may even lead to Alzheimer's disease. The lecithin molecule is technically named phosphatidylcholine because it can be broken down into glycerin, two fatty acids, phosphorous and choline. The choline portion keeps the cell membrane fluid and flexible, allowing for the passage of nutrients into (and wastes out of) the cell.

Some British researchers believe that lecithin repairs the cell membranes of unhealthy liver cells and thus has some antiaging properties. Lecithin may possibly protect against heart disease because it helps to elevate "good cholesterol." It may also improve short-term memory because the choline in the lecithin is used in the biosynthesis of acetylcholine, a neurotransmitter required by many nerve cells.

LEUCINE, an essential branched-chain amino acid, is available in animal protein, wheat germ, and AFA algae. Similar to isoleucine, it is concentrated in muscle tissue and used to produce energy. It stabilizes blood sugar and may be particularly helpful in reducing some of the uncomfortable symptoms of hypoglycemics.

LEUKOCYTES. *See* white blood cells.

LINOLEIC ACID, an essential fatty acid, must be obtained from a good diet. It cannot be biosynthesized by humans and must be continuously supplied to sustain life. Linoleic acid is a precursor for gamma-linolenic acid (GLA), which is required for needed prostaglandin synthesis. However, various factors "block" linoleic acid from being converted to GLA. Saturated fats, cholesterol, alcohol, viral infections, and zinc deficiency act as blocking agents. Thus, even though we may get linoleic acid from our diet, we often do not have the ability to metabolize it to GLA. Consuming AFA algae on a daily basis offers a direct source of GLA.

LIPASE is a digestive enzyme that breaks down lipids, usually in the form of triglycerides, in the small intestine. The enzyme originates in the pancreas and is called pancreatic lipase.

LIPIDS are fats, waxes, and fatty acids found in all plants and animals. They are insoluble in water. Lipids are usually found in the form of triglycerides, a chemical combination of the "sweetener" glycerol with three attached fatty acids that vary in their similarity to each other. Fat digestion takes place in the intestine with the help of the liver's bile (an emulsifier) and pancreatic enzymes. They are broken into fatty acids (and glycerol), which are later resynthesized back into lipid triglycerides. Since lipids make up about 60 percent of the brain, their dietary source ought to be as good as those found in AFA.

LIPOPROTEINS are complex proteins that are chemically bonded to lipids (fat). Low density lipoproteins ("bad" cholesterol) contain about 35 percent cholesterol, whereas high-density lipoproteins ("good" cholesterol) contain about half that much (17 percent).

When white blood cells encounter an LDL molecule damaged by free radicals, they engulf them, become engorged, and then stick to the artery walls as "foam cells." This blocks arteries and causes high blood pressure. AFA may help to reverse this effect.

LIVER, the largest organ in the body, has an extraordinary wide variety of functions. The liver secretes bile which breaks up fat particles in the small intestine. This organ also stores glycogen for later use, along with iron and various B vitamins. All blood returning from the stomach, intestines, spleen, and pancreas is "detoxified" by the liver such that most poisons, drugs, and medicines can be broken down and rendered harmless.

LYMPH is a fluid that flows parallel to the blood circulatory system and primarily contains lymphocytes. The Latin word *lympha* means water.

LYMPHOCYTES are white blood cells that fight infection. When stimulated by an antigen such as foreign proteins or polysaccharides, lymphocytes produce antibody proteins that bind to their surface. They number about a trillion in our bodies and pass through the thymus gland, a walnut-sized organ at the base of the neck. When "fighter cells" called T-lymphocytes (T for thymus) encounter an invading virus, they surround the human cell within which the invading virus has hidden. These T-cells, sometimes called "natural killer" cells (NKs), release toxic molecules that open up these cells, releasing the foreign virus (antigen). The T-cells then release their antibodies which bind to the surface of the protein-coated virus. Fever, swollen lymph glands, and muscle aches usually accompany this process. Now that the virus is "marked" by the bonded antibody, white blood cells called macrophages ("big eater" in Greek) engulf, digest, and destroy the virus. Vitamins B_6 and B_{12} along with amino acids help the T-cells to produce the antibodies that mark invaders. AFA is very rich in all of these immune-stimulating ingredients.

LYSINE is an essential amino acid that is available in wheat germ, pork, fish, dairy products, and AFA. Concentrated in muscle tissue, lysine, along with vitamin B_6, helps in the intestinal absorption of calcium. This helps to prevent osteoporosis. Lysine may also help to reduce lead toxicity by chelation. Along with vitamin E and iron, lysine is used in collagen (skin protein) formation.

LYSOZYME, an enzyme in teardrops, protects the eyes by chemically destroying the cell walls of invading bacteria, rendering them harmless.

MAGNESIUM, an important "anti-stress" mineral, plays a role in about 325 different enzyme systems that control almost all vital biochemical processes. A magnesium deficiency can bring on migraine headaches and high blood pressure; getting enough magnesium will contribute toward a healthy and vibrant feeling. Unfortunately, magnesium deficiency is much too common. The USDA tells us that the average American gets only 25 percent of the RDA of magnesium. Long-term deficiency includes fatigue, twitching, and confusion. The magnesium in AFA may also be helpful in reducing anxiety, irritation, mood swings, and fatigue. The best dietary sources of magnesium are AFA, seafood, greens, nuts, and grains.

The magnesium in AFA is chelated inside of the chlorophyll molecule. As such, it is effective as a natural tranquilizer because it relaxes skeletal muscles and the heart and also reduces fatigue. Since hard water is high in both calcium and magnesium, in geographical areas where the water is hard, people have a lower incidence of cardiovascular problems. The magnesium in AFA may also be helpful in reducing premenstrual stress syndrome symptoms. Magnesium absorption is low partly because sugar and dairy products seems to lessen its absorption. The phosphate in all soft drinks will leech valuable magnesium from the body. Diabetics, the elderly, alcoholics, and pregnant women would all probably benefit from the high concentration of magnesium in AFA.

MALABSORPTION SYNDROME is a general disease in which nutrients cannot be properly absorbed and utilized. Malnutrition is the resulting condition.

MANGANESE is a little-understood trace mineral that is important in the formation of bone and in the metabolism of glucose and protein. Although the human body contains only a "pinch" (15–20 milligrams) of this metal, it is enough to catalyze the biosynthesis of dopamine, an important neurotransmitter. Manganese is found in small amounts in greens, nuts, seeds, whole grains, and AFA. Unfortunately, high-tech farming depletes manganese such that most people are deficient in this mineral. Refinement of wheat to white flour depletes 90 percent of its manganese.

Manganese probably helps to decrease bone joint inflammation by increasing the production of soothing mucopolysaccharides. This is partly related to the role manganese plays in a variety of antioxidant enzyme systems. It may be helpful in relieving fatigue, nervousness, and poor memory. Since cancerous tumors are known to be exceptionally low in manganese, this mysterious mineral may play an unknown anti-cancer role.

MANNA was a food referred to in the Bible and eaten by the wandering Hebrews in Israel. Some archaeologists speculate that it was a blue-green algae, which dried on rocks and turned into a sweet white polysaccharide.

MEDICAL CHECK-UPS are important every year for a variety of reasons. Ask your doctor about getting a "biochemical analysis" of both blood and urine. These evaluations can yield valuable data.

METABOLISM refers to all biochemical reactions which involves either the synthesis (anabolism) or breakdown (catabolism) of molecules inside any cell.

METHIONINE is an essential amino acid found in nuts, sunflower seeds, rice, corn, liver, eggs, fish, and AFA algae. It contains sulfur (as does cysteine) and is therefore important for skin and nail growth, fat reduction, and preventing fatigue, depression, and some allergies. When bioconverted to cystine, methionine also serves as an antioxidant. Some studies suggest that "excessive" alcohol destroys or otherwise greatly reduces the amount of methionine.

Methionine biostimulates specific enzymes with the help of vitamins B_{12} and B_6. It can then increase the amount of memory-enhancing neurotransmitters such as dopamine, and mood-elevating polypeptides such as endorphins. These polypeptides can properly be absorbed from AFA directly into the bloodstream, thus positively affecting physiological behavior directly. Methionine was probably one of the first amino acids available in earth's ancient primordial seas, billions of years ago. This amino acid was (and is still) used by blue-green algae to biosynthesize glutathione, Earth's first antioxidant. Methionine has also been shown to help humans detoxify lead and copper in the blood. Studies have even shown radiation protection as well. Some of our friendly bacteria (if we have healthy intestinal flora) can synthesize this needed methionine from the more accessible amino acid, aspartic acid.

MICROGRAM is a unit of weight equal to one-millionth of a gram. It is abbreviated as *mcg*.

MICRONUTRIENTS are profoundly necessary physiological substances such as trace minerals, vitamins, certain essential amino acids, coenzymes, antioxidants, bioflavonoids, and essential fatty acids present in and required by the body with a concentration typically measured in micrograms or milligrams.

MINERAL ASH (OR ALKALINE ASH) refers to the residue of a food after its complete combustion or burning with oxygen gas. Mineral compounds such as magnesium oxide and calcium oxide are left over and weighed as a measure of alkalinity.

MINERALS are rocks and ores that contain metallic elements (e.g., iron, calcium, and zinc) and nonmetal elements (e.g., chloride and phosphate) that are absolutely essential for the existence of all life forms – plant, animal, bacteria, or algae. Minerals such as calcium phosphate are used by the body in the form of bones and teeth; minerals such as iron or zinc are essential components for the glorious workings of enzymes.

MITOCHONDRIA are regions within eukaryotic cells where energy is stored as adenosine triphosphate, or ATP. Each mitochondrial "organelle" is a magnificent factory of "oxidative enzymes" that slowly release enough energy so that the cell can survive. Ninety percent of the molecular oxygen used by the cell is used up in the mitochondria.

MOLECULE is the smallest unit of a chemical compound capable of existing independently while retaining properties of the original substance.

MOLECULAR SHIELDS are protective antioxidant molecules that can disarm dangerous forms of oxygen called free radicals, which are generated during many stages of cellular metabolism. The most important antioxidant molecular shields in AFA are betacarotene, superoxide dismutase, glutathione, vitamin C, and vitamin E.

MOLYBDENUM is a very rare mineral and important micronutrient. It is available in AFA algae and can bioactivate several human enzymes associated with longevity enhancement, free-radical absorption, and the ability to neutralize carcinogenic nitrosamine molecules, which are linked to colon cancer.

MONOCYTES are large leukocytes (white blood cells) that can attack and engulf great numbers of invading bacteria, destroying them with the aid of powerful enzymes.

MONOSACCHARIDES are simple sugars composed of single molecules. AFA contains glucose, galactose, mannose, and ribose monosaccharides.

NEURONS are specialized nerve cells that can communicate with each other by long cellular extensions that conduct electrical impulses. The flexibility of neurons is enhanced by high-quality essential fatty acids, such as those found in AFA.

NEUROTRANSMITTERS are molecules that transmit messages through nerve tissue. They are usually composed of several amino acids and play a key role in mental states such as anxiety and depression. The interaction of neuropeptides and receptor sites are where images turn into chemistry, chemistry into images.

NIACIN (VITAMIN B3) is the "stress reducer" B vitamin water-soluble molecule. It is readily obtained from AFA, meat, fish, and green vegetables and is used in enzymes to repair cells and convert food into energy. Niacin actually refers to either nicotinic acid or nicotinamide. A lot of research points to a marked cholesterol-lowering effect by niacin, even in relatively small amounts. In the form of nicotinic acid, it may be able to reverse some atherosclerosis. Blood triglycerides have been lowered by as much as one-half in some human studies, followed by blood pressure lowering. Niacin often detoxifies damaged liver cells in alcoholics and heroin addicts and thus lowers cravings for these substances.

NICKEL is thought to be an essential trace mineral and is involved in bioactivating certain enzymes that break down glucose. Deficiency of this mineral may impair the liver. Nickel is found in soybeans, green peas, and AFA algae.

NITROGEN FIXATION is carried out in the heterocyst cells of blue-green algae. With the help of nitrogenase enzymes, nitrogen gas can be transformed into ammonia for the eventual manufacture of amino acids.

NUCLEIC ACIDS are polymers of ribose sugar and phosphate, alternating with purine and pyrimidine molecules (nitrogenous organic bases), that are found in all cells except human red blood cells. The two major types of nucleic acids are DNA and RNA. DNA (found in the nucleus) and RNA (outside the nucleus) work together to store genetic information that dictates how all proteins and enzymes can be correctly synthesized.

NUTRIENTS are biochemical substances needed for the fundamental survival of life. Nutrients are made available to us as proteins, carbohydrates, lipids (fats), vitamins, minerals, and water.

NUTRITION is a constant and daily process of being nourished. This can refer to food, water, or air, as well as all else that enters our body and mind, such as our daily thoughts, sights, and psychological experiences. Good nutrition also means ingesting those nutrients which – by optimum digestion and nutrition – positively affect the health of every cell of our body. This makes nutrition an important aspect of preventive medicine as well as an often overlooked component in corrective medicine.

"There is no better food than meeting a friend face-to-face."
– *W. B. YEATS*

NUTRITIONAL STATUS is a qualitative measure of the well-being of the body as observed by the overall condition of the eyes and skin, tongue, and gums.

OLIGOSACCHARIDES are monosaccharide polymers found within the cell walls of blue-green algae.

Organelles are substructures or "little organs" within the cell such as chloroplasts, mitochondria, and ribosomes. Some biologists speculate that organelles were once independent cells. The photosynthetic chloroplasts that give plants their green color may have originally been primitive, yet independent forms of blue-green algae.

Organic is a word with two meanings. To the chemist it simply means molecules that contain carbon (other than CO_2 and metal carbonates). To the consummate organic farmer, it refers to farming practices that do not employ the use of pesticides and herbicides.

Organic acids are molecules that contain one small cluster of carbon, oxygen, and hydrogen atoms (called a functional group) usually placed at one end of the molecule. When placed in water, the hydrogen atom is pulled off as an ion, and the pH is lowered. Organic fatty acids such as linoleic or linolenic acids are well-known examples of long-chain acids; acetic and lactic acids are examples of short-chain acids.

Orthomolecular medicine is a medical treatment that uses the body's own naturally occurring substances (e.g., amino acids) to promote nutritional balance. The word "ortho" simply means normal or endogenous to human physiology. According to the late Dr. Linus Pauling – who originated the term – a natural diet supplemented with orthomoleculars is essential for good health. AFA provides just such a vital spectrum of orthomolecular nutrients.

Osteoporosis is a disease whereby bones become porous due to insufficient bone protein production.

Oxidation refers to a general type of chemical reaction, which involves either the addition of oxygen atoms, the removal of hydrogen atoms, or both. For example, when glucose is oxidized by molecular oxygen, glucose's carbon and hydrogen atoms both break away and become enriched with oxygen. When a saturated fatty acid is oxidized, unsaturation develops as hydrogen atoms are removed. In general (and from a more modern perspective), carbon in food undergoes oxidation by *losing electrons* during the Krebs cycle within the cell's mitochondria.

Pancreas is an organ (behind the stomach) that secretes pancreatic enzymes, which digest a variety of starches, fats, and proteins. In addition, the hormones insulin and glucagon are produced to regulate glucose levels.

Pancreatin is a secretion from the pancreas that contains digestive enzymes (e.g., amylase, trypsin, and lipase).

Pancreozymin, a hormone produced by the wall of the small intestine, stimulates the production of pancreatic "juices." These juices contain various types of digestive enzymes such as proteases (to digest proteins), lipases (to digest fats), amylases (to digest carbohydrates), and nucleases (to digest nucleic acids).

Pantothenic acid (vitamin B5), the "anti-aging" vitamin, plays a variety of roles in enzymatic reactions throughout the body. It is obtained from liver, eggs, AFA algae, and whole grains. This vitamin is especially important for how your body uses amino acids to form proteins and manufacture hormones. Vitamin B_5 is also an antioxidant that helps to protect cells from the damage of free radicals. It alleviates some morning arthritis pain and joint disorders because of such antioxidant properties. It also biostimulates the immune system by activating certain types of white blood cells. One of the B_5 enzymes may even speed up the detoxification of alcohol toxins (acetaldehyde) in the blood.

PAPAIN is a protein-digesting (proteolytic) enzyme from papaya.

PASTEURIZATION is a food process that uses mild heat to reduce microbial levels in substances, such as milk, which are sensitive to heat. Endogenous enzymes are usually destroyed in this process.

PHAGOCYTES refer to specialized immune cells (macrophages) in the blood and lymph fluid that are capable of traveling freely in body fluids to ingest bacterial invaders. Since their ability to do so depends partly on the flexibility of their cell membranes, ingesting essential fatty acids – such as those in AFA – is helpful.

PHARMACOLOGY is the study of the action of drugs on the body. Drugs will only increase or decrease a specific action already present in a cell. Drugs will *not* introduce a new action in the body as will the biostimulating effects of nutrient-dense foods such as wheat germ, alfalfa sprouts, and AFA.

PHOSPHOROUS is a nonmetal mineral element that is present in every cell, especially in bones and teeth. It is vital to energy production and protein synthesis and is readily available from most protein foods such as meat and AFA algae.

PEPTIDE BONDS are those which hold one amino acid to another in a peptide or protein. Such as bond is sometimes called an "amide linkage," and digestion is needed, with an appropriate enzyme, to break it.

PEPTIDOGLYCAN is the rigid layer of a cyanobacterial (blue-green algae) cell wall that is composed of several types of complex sugars and short polypeptides.

PESTICIDES are chemicals used to destroy insects and animals that impact agriculture. Currently the Environmental Protection Agency allows more than 400 varieties of pesticide, equaling 2.6 billion pounds, to be used per year in the United States. This comes out to approximately 10 pounds per year for every man, woman, and child in America. When laboratory animals are given three pesticides (instead of an equal weight of one), death usually results. This is called "multiple exposure syndrome." (*See* Appendix B)

pH is a measure of the acidity (or basicity) on a scale ranging from 0 to 14. The letter "p" stands for "power" and the letter "H" stands for hydrogen ion concentration. A pH value less than 7 is acidic, more than 7 is basic, and equal to 7 is neutral. The pH of blood is basic 7.4.

PHENYLALANINE is an essential amino acid found in milk, meat, wheat germ, and AFA. This interesting molecule has a direct effect on brain chemistry because of its relatively rare ability to very quickly cross the "blood-brain barrier." In the liver, phenylalanine is easily bioconverted to such important neurotransmitters as adrenaline and dopamine.

With the help of B vitamins, phenylalanine enhances alertness, concentration, and learning abilities. Phenylalanine seems to diminish addictive cravings for alcohol and sugar. By slowing down the breakdown of morphine-like enkephalin and endorphin peptides, back pain, arthritic pain, and even depression are reduced. The phenylalanine in AFA is also useful in a wide variety of brain and intestinal peptides. Long term use of smaller amounts (as in AFA) has not yet been carefully studied. Unfortunately, drinking diet sodas that contain the artificial sweetener aspartame, a dipeptide of phenylalanine and aspartic acid, can cause excessive absorption by the brain and diminish some of the beneficial effects of AFA.

PHOSPHOLIPIDS are triglyceride fat molecules that help to form flexible cell membranes.

PHOSPHATIDYLCHOLINE is the most common phospholipid present in and important to the structure and flexibility of all cellular membranes – plant, animal, or algae. It consists of phosphate, choline, and a B vitamin.

PHOTOSYNTHESIS is a process whereby plants and algae utilize the energy of the sun and the catalytic presence of chlorophyll to combine CO_2 and H_2O to synthesize glucose. Carried out by a complex chain of chemical events, all life in Earth's biosphere ultimately depends upon this transformation. Actually about *HALF OF ALL PHOTOSYNTHESIS ON THIS PLANET IS CARRIED OUT BY BLUE-GREEN ALGAE,* one of the earliest pioneers of this life-giving process.

AFA carries on photosynthesis within ancient biofactories called thylakoid sacs, which are complex betacarotene- and chlorophyll-containing factories that assimilate sunlight. Their photosynthetic efficiency is reflected by the fact that no other life form has such generous amounts of these transformative pigments.

PHYCOERYTHRIN is an antioxidant pigment that can give blue-green algae a red coloration.

PHYCOLOGY is the study of algae.

PHYTOPLANKTON are suspended algae that drift freely with the water's current. The great blue whale subsists solely on such algae.

PIGMENTS are colored molecules. Important examples are green chlorophyll, orange-red carotenoids, and red-blue anthocyanins.

PLAQUE found on artery walls mostly contains cholesterol and triglycerides. It contributes to artery "hardening" and partial artery blockage. This typically results in high blood pressure or even the complete blockage of arteries.

PLASMA is the liquid that carries blood cells, nutrients, gases, wastes, hormones, and antibodies. When blood cells are removed from plasma, blood serum remains.

PLATELETS are a type of white blood cell that causes blood clotting and, after liberating serotonin, causes small blood vessels to severely contract to prevent further blood loss.

POLYPHOSPHATE GRANULES, found inside the AFA cell, are composed of phosphate polymer chains of phosphorous and oxygen, and are used to construct AFA's cell membrane.

Polysaccharides are polymer chains of simple sugars that form starch, cellulose, and glycogen.

Polyunsaturated refers to organic molecules that have one or more double bonds between one or more pairs of bonded carbon atoms. Polyunsaturated fatty acids in AFA algae help to keep cell membranes fluid and flexible.

Prebiotic soup refers to the physical and chemical conditions on planet Earth right before the beginning of life and the proliferation of one-celled alga organisms.

Precursors are biomolecules that can be synthetically converted to another useful form. For example, the betacarotene of AFA is a precursor to vitamin A.

Prokaryotic cells, such as bacteria and blue-green algae (cyanobacteria), do not have a nucleus or nuclear membrane.

Prostaglandins are powerful hormone-like molecules that affect inflammation, muscle contraction, cholesterol levels, and blood pressure. The essential fatty acids of AFA help to synthesize those prostaglandin hormones that are beneficial in raising HDL ("good" cholesterol).

Probiotics is a Greek word meaning "for life" and refers to the friendly bacteria (about 400 different species) that are symbiotically needed in the gastrointestinal tract to maintain good health. They improve digestion and profoundly increase proper bowel function. Typically, they change the acidity of their surroundings and thus kill invading, unfriendly toxic bacteria.

Proline, a non-essential amino acid, is a component of collagen and thus helpful in tendon and cartilage formation. Readily found in meat, dairy, wheat germ, and AFA, proline peptides also contribute to enhanced learning abilities.

Prostaglandins are extremely potent fatty acids that act like regulator hormones to produce neurochemical effects in very small amounts. They are vital enzyme regulators and hormones that control every cell in every organ.

Protease is a protein-digesting enzyme, which includes pepsin, trypsin, chymotrypsin, and peptidases.

Proteins are polypeptide macromolecules made up of a linear arrangement of twenty participating amino acids. The sequence and number of amino acids affect the shape and function of the resulting protein. Some are catalysts (enzymes), and others are structural agents inside and outside the cell. AFA algae is about 60 percent protein with all eight of its essential amino acids in perfect balance. Beef, by comparison, is only about 20 to 25 percent protein.

Ptyalin is a starch-digesting amylase enzyme found in the saliva and released by chewing.

Putrefaction refers to the decomposition of proteins in the small intestine, which gives off a foul smell or bad breath.

Pyridoxine (vitamin B_6) is the "immune system booster" vitamin that is obtained primarily from whole grains, brewer's yeast, and AFA. Because this vitamin bioactivates more than sixty enzyme catalysts, it plays a key role in a wide variety of biochemical reactions such as the production of immune and red blood cells as well as the biosynthesis of neurotransmitters.

RDA means "recommended daily allowance" in reference to vitamins, minerals, and other micronutrients. Because of poor diets, most people do not reach these allowances. Most older Americans, for example, receive only about two-thirds of their B vitamins and even less of the minerals they need. The RDA of nutrients are based on the standards established by the Food and Nutrition Board of the National Research Council.

"The RDA for a vitamin is not the allowance that leads to the best of health for most people. It is, instead, only the estimated amount that for most people would prevent death or serious illness from overt vitamin deficiency."

– LINUS PAULING *(two-time Nobel Prize winner)*

RECEPTOR SITES are regions on the surface of cell membrane proteins that allow messenger molecules (e.g., hormones, neurotransmitters) to temporarily connect so that important chemical reactions may commence.

RED ALGAE, or rhodophyta (from the Greek *rhodos*, red), are a type of "seaweed" that carry on photosynthesis deep within the sea. They are most abundant in warm tropical coastal waters and have been found at depths of 800 feet near the Bahamas. They have a limited use as a harvested food source because of their cellulose cell wall.

RED BLOOD CELLS (ERYTHROCYTES) number about 25 billion in the body. They live about 4 months while being bent, squeezed and deformed as they pass through our tiny capillaries. Creation of a new red blood cell takes about a week in the bone marrow with the help of vitamin B_{12}. Red blood cells do not undergo cell division. Instead, about 1.2 billion red blood cells die every hour and need replacement. AFA is unsurpassed as a source of vitamin B_{12}.

RHAMNOSE SUGARS are complex natural plant sugars found in algae.

RIBOFLAVIN (VITAMIN B2) is an "antioxidant" water-soluble, B-vitamin molecule that helps to biosynthesize glutathione, an important antioxidant that protects against free radical damage. Riboflavin is obtained from AFA, milk, and green vegetables. Riboflavin deficiencies are very common, especially in the elderly and alcoholics. Cracks on lips or in the corners of the mouth show this deficiency.

RIBOSOMES are cellular particles in all cells made up of proteins and ribonucleic acid (RNA). Their function is to accurately transcribe the genetic code of DNA and manufacture peptides and proteins from amino acids.

RNA, OR RIBONUCLEIC ACID, is a polymer of ribose sugar, phosphate, and, most critically, the nucleic acids that are intimately involved with DNA in the synthesis of proteins. *Aphanizomenon flos-aquae* is rich in RNA, the nucleic acids of which help regenerate the "mind-brain" function and reportedly enhance memory, concentration, and mental clarity.

SATURATED FATTY ACIDS are derived from animal and vegetable fats. Dietary sources of these molecules should be avoided because they impart rigidity to the cell membrane. They are composed of long-chain organic acids, the carbon atoms of which have the maximum (saturated) number of bonded hydrogens.

SELENIUM is an antioxidant and an immuno-stimulating mineral that also provides valuable protection against certain forms of cancer. It is found in wheat germ, liver, and AFA. Selenium concentration varies dramatically with the soil from which it is derived. For this reason, garlic and onions vary widely in selenium content.

SERINE is a nonessential amino acid used primarily in brain proteins and nerve coverings. It is used in DNA and RNA synthesis and is especially important in the formation of the phospholipids needed for flexible cell membranes. Found in meat, dairy, and AFA, serine is good for the skin and has often been used as a natural moisturizer in skin creams.

SEROTONIN is a powerful neurotransmitter. It is biosynthesized from the amino acid tryptophan found in AFA. Serotonin is a brain hormone that can reduce depression, elevate mood, constrict blood vessels, regulate sleep, trigger intestinal peristalsis, and change into melatonin within the pineal gland.

SPECIES are a collection of closely related strains.

SPLEEN is an "organ" that is actually a collection of lymph tissue and serves as a reservoir for red blood cells.

STANDARD AMERICAN DIET (S.A.D.) usually consists of processed fats, refined flour and sugar, and too many food additives and pesticides. Most of the U.S. population receives only about three-quarters of their needed vitamins and minerals. The reason for this is that most of the U.S. food supply is grown in nutritionally depleted soil. This leads to an alarming variety of diseases such as heart disease, diabetes, schizophrenia, arthritis, and cancer. Early warning signs of chronic vitamin and mineral deficiency, such as fatigue, mood swings, confusion, anemia, and insomnia, affect the quality of life, and immune deficiencies affect our very life span. Sensible prison officials have discovered that by simply removing refined flour and sugar from the meals served to their prisoners, there was a dramatic reduction in the amount of violent behavior.

STARCH is a glucose polymer found and stored in plant seeds and roots.

STRAINS are a population of cells descended from a single cell of a particular species. There are seven different aphanizomenon species and ten different strains of AFA.

STRESS is a general term that describes a biochemical tension – be it psychological, environmental, or physiological – to which the body is subjected. Since micronutrients are expended in dealing with stress, micronutrients need to be continually replenished to avoid the weakening of the entire immune system.

SUBSTRATE MOLECULES are nutrients that are targeted and transformed into cellular components by the catalytic action of biochemical enzymes.

Superoxide dismutase (SOD) is a significant enzyme and a ubiquitous antioxidant that protects important biomolecules such as DNA from being damaged by naturally generated superoxide free radicals. All organisms contain SOD to varying extents, AFA more than most.

Substance P is a powerful neurotransmitter composed of interconnected amino acids, which are available in AFA. The amazing effect of this neuropeptide is that it "sharpens" the mind by stimulating brain cells to grow more dendritic spines that can enhance learning ability and elevate our sense of self-worth.

Synthesis is the combining of constituents into a unified whole. In biochemistry it refers to the combining of atoms, ions, and/or molecules for the purpose of creating or manufacturing new substances. When new substances are synthesized in the body, enzymes are almost always used.

Synergy is the energetic interaction of various nutrients so that their combined effect is greater than the sum of their individual effects.

Thiamine (vitamin B_1) is the "nerve and energy" B-vitamin molecule (water-soluble) that is involved in numerous nerve, heart, and muscle tissue reactions. It is obtained from AFA algae, brown rice, and grains, and helps to convert blood glucose into energy by activating the necessary enzymes. Deficiency of vitamin B_1 produces fatigue and mental confusion (beriberi disease). Alcohol can produce this deficiency by chemically impairing and reducing this or any B vitamin's absorption in the intestines. Heavy tea and coffee drinkers and the elderly may also have this problem. Vitamin B_1 seems to prevent lead – a ubiquitous pollutant- from depositing in nerve and brain tissue. The biologically active and chelated form of vitamin B_1 (thiamin carboxylase) is easily absorbed and assimilated. The vitamin B_1 in AFA is chelated in a similar manner, allowing even a small amount to have beneficial effects. Vitamin B_1 is helpful toward alleviating nervous system disorders. It seems to improve mental clarity. A number of factors in AFA do this. The synergy of all of them might help to explain why so many people using AFA report a dramatic mental upliftment.

Thiaminase is an enzyme present in some fish and thought to be produced by some of the friendly intestinal bacteria of humans.

Threonine, often called the "immune booster," is an essential amino acid found in dairy, meat, wheat germs, nuts, seeds, and AFA. It is needed especially for skin proteins such as collagen and elastin and plays a minor role in reducing fat buildup in the liver. There is some evidence that threonine can promote healing by biostimulating the thymus gland to produce T-cells with more flexible cell membranes.

Thrombin is an enzyme found in blood that helps bring about blood clotting. Vitamin K, produced by friendly bacteria, is used in the biosynthesis of this enzyme.

Thylakoids are photosynthetic chlorophyll-containing membranes in cyanobacteria (blue-green algae).

THYROID GLAND is an endocrine gland located in the front of the neck, behind the larynx. It produces thyroxin hormone, an important blood-circulating molecule that regulates cellular energy production. Thyroxin also controls glucose absorption in the intestine and glucose metabolism in each cell. The cells in this "master" gland need iodine to biosynthesize this hormone.

TOXIC BOWEL is a common condition in which toxins from bowel bacteria are present to cause diabetes, ulcerative colitis, thyroid conditions, and assimilation and digestive problems. Fiber present in AFA tends to bind to bacterial toxins in the bowel and get excreted. Along with a good probiotic regimen that includes *acidophilus* bacteria, AFA algae's assortment of complex carbohydrates, balanced profile of amino acids, and wide range of minerals and vitamins (especially betacarotene) are therapeutically significant to reverse such symptoms.

TRACE ELEMENTS are micronutrients such as selenium and chromium that are required by humans in very small quantities. These micronutrient trace elements are typically utilized as enzyme activators. AFA offers the widest range of such trace elements – *more than any known food.*

TRANS-FATTY ACIDS are unsaturated chains of carbon atoms, ending in an organic acid group. The "trans" variation usually contributes to an inflexible cell membrane. This negative effect is due to the relatively straight shape of this molecule.

TRANSAMINASE enzymes are those that allow for the chemical transfer of an amino group from one amino acid molecule to another. Vitamin B_6 is required as a coenzyme for this transamination process. Amino acids are termed "essential" when there are no transaminase enzymes available to make them. AFA has all the transaminase enzymes it needs to make the amino acids that humans require.

TRYPSIN, a protein-digesting enzyme, disassembles partially digested proteins in the small intestine. It is so specific that it only breaks the bond between lysine and arginine.

TRYPTOPHAN, an essential amino acid, may be considered an essential vitamin because it is used in the biosynthesis of niacin, or vitamin B_3. Tryptophan is a precursor for the very important neurotransmitter serotonin, a brain hormone and general messenger molecule needed by the brain to bring on sleep and affect mood patterns. Tryptophan is found in small amounts in AFA, turkey, and milk, as well as some nuts and seeds.

TYROSINE is a nonessential amino acid that is concentrated mostly in muscle tissue and a little in the brain. It is biosynthesized from and biochemically similar to phenylalanine, an essential amino acid. Known as the "anti-depressant" amino acid because of its importance to brain nutrition, it promotes mental alertness, learning, and memory.

UPPER KLAMATH LAKE is the largest lake in Oregon and the home of *aphanizomenon flos aquae* – AFA blue-green algae. Luckily, the lake averages only eight feet deep, thousands of years of mineral runoffs and about 300 days of sunshine per year help to create the ideal conditions needed for our precious algae blooms.

Valine is an essential branched-chain amino acid. Charitably enough, it is available in most foods and builds muscle tissue. A valine deficiency has produced "aimless circling" in laboratory animals. This could be caused if too much isoleucine and leucine go to the brain, blocking transport of valine. Amazingly, the proportions of all three of these interrelated BCAAs in AFA is nearly identical to the proportions recommended by the National Academy of Sciences.

Vanadium is an essential trace metal slightly present in good soil. It is found in unsaturated olive oil, carrots, radishes, cabbage, and AFA. It is especially concentrated in black pepper and dill seed. Vanadium is involved in enzyme systems that regulate blood sugar, bone formation, fat metabolism, and red blood cell production. There was a time when vanadium was considered a cure-all by turn-of-the-century French physicians. Still today there is much that is unknown and more research is needed to verify its biochemical functions.

Viruses are living bundles of protein and nucleic acids capable of destroying the host cells within which they reproduce, especially in weakened immune systems.

Vitamin A is an antioxidant vitamin that is stored in animal livers and available from plants and blue-green algae in the form of betacarotene.

Vitamin B1. *See* thiamine.

Vitamin B2. *See* riboflavin.

Vitamin B5. *See* pantothenic acid.

Vitamin B3. *See* niacin.

Vitamin B6. *See* pyridoxine.

Vitamin B12. *See* cobalamine.

Vitamin C, or ascorbic acid, is an antioxidant and an immuno-stimulating vitamin that helps produce collagen and aids in the absorption of iron. It is somewhat effective in reducing the duration of the common cold by about 30 to 50 percent, deters gum bleeding, maintains good vision, and may be helpful in preventing diabetes. Research evidence shows that a daily intake of this vitamin may reduce the incidence of stomach and esophagus cancer.

Vitamin E (tocopherol) is a "nerve healing" vitamin that is essential to animals and is an important antioxidant (free radical protector), as is betacarotene. It is widely distributed and concentrated in vegetable oils, whole grains, eggs, and AFA. Unfortunately, half the amount of ingested vitamin E is excreted. Tocopherols (*tocopherol* is a Greek word meaning "body bearer") help to prevent the oxidation of other fat-soluble vitamins. This vitamin, by strengthening cell membranes, keeps invading viruses from penetrating into a cell. Also, red blood cells appear more healthy and have prolonged lives. It is now an established fact that vitamin E will slow down or even reverse a variety of neurological disorders caused by nerve damage.

Vitamin F. *See* essential fatty acids.

VITAMIN K, OR PHYLOQUINONE, is a fat-soluble micronutrient produced by friendly intestinal bacteria and associated with the chlorophyll of green plants and algae. It is sometimes called the "anti-hemorrhage" vitamin. In fact, the letter K stands for the German word *koagulation*. Deficiency is uncommon so long as friendly intestinal bacteria continue to manufacture this mysterious tumor-inhibiting substance. People with dysfunctional bowel syndrome have a dangerously-low probiotic population and risk vitamin K deficiency. Bruising easily and developing black-and-blue marks is a warning of this deficiency. Taking *acidophilis* to nourish the microflora is strongly recommended.

VITAMINS are biomolecules that are required for health and survival. Because English scientist, Casimir Funk, showed in 1906 that these molecules were "vital," and thought they belonged to the "amine" family, he gave vitamins their name. Although they are vital, they are now known to extend well beyond the amine family. Sometimes essential fatty acids are collectively referred to as vitamin F.

WHITE BLOOD CELLS (LEUKOCYTES) include many different cell types with varying functions and purposes. Roughly, for every 1000 red blood cells there is one white blood cell. White blood cells live approximately 12 days, while red blood cells live from 100 to 120 days. There is a rise in the white blood cell population after an infection. White blood cells then destroy invading viral and bacterial particles with an "amoebic-like" action. Even oxidized fats left over from partially digested meals are engulfed. White blood cells use their supple and graceful motion to actually pass through capillary walls and enter infected tissues, as needed.

ZINC is a profoundly important micronutrient mineral because of the role it plays in more than 100 different enzymes. It can build healthy cell membranes, maintain healthy skin, biostimulate the immune system, and help in the synthesis of DNA and RNA. One zinc-containing enzyme (carboxypeptidase) improves protein digestion. Zinc is found readily chelated in AFA algae, oysters, liver, and pumpkin seeds. It can help to reverse the gradual breakdown of an aging immune system and also prevent and treat the common cold. Stored in the pancreas, zinc reduces some of the symptoms of diabetes and improves the health of the prostate gland. Zinc can also act as an antioxidant and anticancer mineral when associated with SOD (superoxide dismutase). Americans get only a fraction of the zinc they need because much of U.S. farmland has been de-zincdefiled by the agro-industry.

Quick Reference Guide

TO THE

Healing Properties

OF

Nutritional Blue-Green Algae

ALANINE	*activates muscles* *boosts the immune system*
ARGININE	*builds new muscles* *detoxifies the liver* *boosts the immune system*
ASPARAGINE	*gives energy to the brain* *builds the immune system*
ASPARTIC ACID	*helps make healthy DNA* *builds the immune system*
BETACAROTENE	*improves vision* *protects the cornea* *helps digestion* *defends against free radical attack* *boosts the immune system*
BIOFLAVINOIDS	*removes toxins from skin cells* *builds the immune system*

Biotin	*produces healthy-looking hair*
Boron	*strengthens bones* *enhances mental clarity* *unclogs arteries*
Calcium	*strengthens bones* *calms nerves* *lowers cholesterol*
Chlorophyll	*promotes bowel regularity* *cleanses interstitial tissues*
Chromium	*moderates existent diabetes* *prevents adult-onset diabetes*
Cobalt	*repairs nerve cell* *helps produce red blood cells*
Copper	*eases arthritis* *helps produce red blood cells*
Cysteine	*detoxifies carcinogens*
Essential fatty acids	*reduces cardiovascular disease*
Electrolytes	*helps kidneys regain optimum function*
Fiber	*eliminates toxic wastes*
Fluorine	*fights tooth decay* *hardens bone*
Folic acid	*improves mental function* *prevents anemia*
Germanium	*helps control Epstein-Barr virus*
Gamma-linolenic acid (GLA)	*lowers cholesterol*
Glutamic acid	*reduces alcohol/sugar cravings*
Glutamine	*improves concentration*

GLUTATHIONE	*defends against free radical attack*
GLYCINE	*calms nervous system*
GLYCOGEN	*increases physical stamina*
HISTIDINE	*enhances nutrient absorption* *removes toxic metals*
IODINE	*regulates body weight*
IRON	*combats fatigue* *lessens depression* *decreases anemia*
ISOLEUCINE	*builds muscle* *helps repair the liver*
LECITHIN	*dissolves cholesterol deposits* *improves short-term memory*
LEUCINE	*reduces hypoglycemic symptoms*
LINOLEIC ACID	*combats viral infections*
LYSINE	*helps prevent osteoporosis*
MAGNESIUM	*promotes tranquillity* *moderates mood swings* *reduces migraine headaches*
MANGANESE	*assists joint mobility*
METHIONINE	*enhances memory* *elevates mood* *removes heavy metals*
MOLYBDENUM	*increases longevity*
NIACIN (VITAMIN B3)	*reduces stress* *lowers cholesterol* *helps to reverse atherosclerosis*

NICKEL	*promotes cellular growth and reproduction*
NUCLEIC ACIDS	*improves memory*
OMEGA-3 FATTY ACIDS	*increases cell membrane flexibility* *dissolves cholesterol deposits* *reduces cardiovascular diseases*
OMEGA-6 FATTY ACIDS	*reduces arthritis symptoms* *improves skin tone* *reduces cardiovascular diseases*
PANTOTHENIC ACID (VITAMIN B5)	*reduces morning arthritis pain* *reduces alcohol toxicity* *defends against free radical attack*
PHENYLALANINE	*increases mental alertness* *reduces sugar cravings*
PHOSPHOROUS	*keeps teeth healthy* *helps repair bone fractures*
POTASSIUM	*reduces hypertension* *helps prevent high blood pressure*
PROLINE	*raises learning ability* *helps repair torn cartilage*
PYRIDOXINE (VITAMIN B6)	*boosts the immune system* *relieves premenstrual tension*
RIBOFLAVIN (VITAMIN B2)	*defends against free radical attack* *provides physical energy* *alleviates eye fatigue*
RNA	*enhances "mind-brain" function*
SELENIUM	*boosts the immune system* *relieves anxiety*
SERINE	*beautifies skin*

Silicon	*tightens the skin*
Sodium	*prevents sunstroke*
Superoxide dismutase (SOD)	*defends against free radical attack*
Substance P	*sharpens the mind*
Thiamine (vitamin B1)	*reduces fatigue* *improves mental attitude* *relieves tension*
Threonine	*boosts the immune system* *ameliorates skin tone*
Tyrosine	*acts as an antidepressant* *promotes mental alertness* *strengthens memory*
Valine	*builds muscle tissue*
Vanadium	*balances blood sugar*
Vitamin B12	*energizes the body* *improves memory* *repairs the nervous system*
Vitamin C	*reduces duration of common cold* *reduces cancer risk* *deters gum bleeding*
Vitamin E	*boosts the immune system* *keeps nerve tissue healthy*
Zinc	*eases memory access* *lessens acne outbreaks* *boosts the immune system* *reduces symptoms of common cold* *helps avoid prostate problems*

ABOUT THE AUTHOR

Karl J. Abrams

A TENURED PROFESSOR OF CHEMISTRY AT SADDLEBACK College in Orange County, California, Karl J. Abrams has taught chemistry for more than twenty years, and is well known for his ability to simplify and make understandable otherwise difficult concepts of chemistry.

Professor Abrams' students and colleagues have praised him for his gift of teaching. He is animated and articulate. He is the author of the successful chemistry textbooks, *A Course in Experimental Chemistry, Books I and II,* published by Freeman Cooper & Co., *Reaction Database Searching Methods* published by Molecular Design Ltd., and numerous articles on chemical education. While studying chemistry as an undergraduate, Mr. Abrams did research under the tutelage of Joshua Lederberg, Nobel Laureate, at Stanford Medical School. After earning his Master's degree from the University of California, he began teaching college chemistry. He spent two years traveling throughout the country lecturing on drug synthesis techniques for that industry. Since then he has focused on investigating and uncovering nutritional research that explains the powerful healing properties of *aphanizomenon flos-aquae* from Upper Klamath Lake, Oregon.

Professor Abrams is a leading authority on nutritional blue-green algae. His lively lectures, stimulating seminars and wonderous workshops are entertaining and educational. He may be contacted through Logan House Publications.

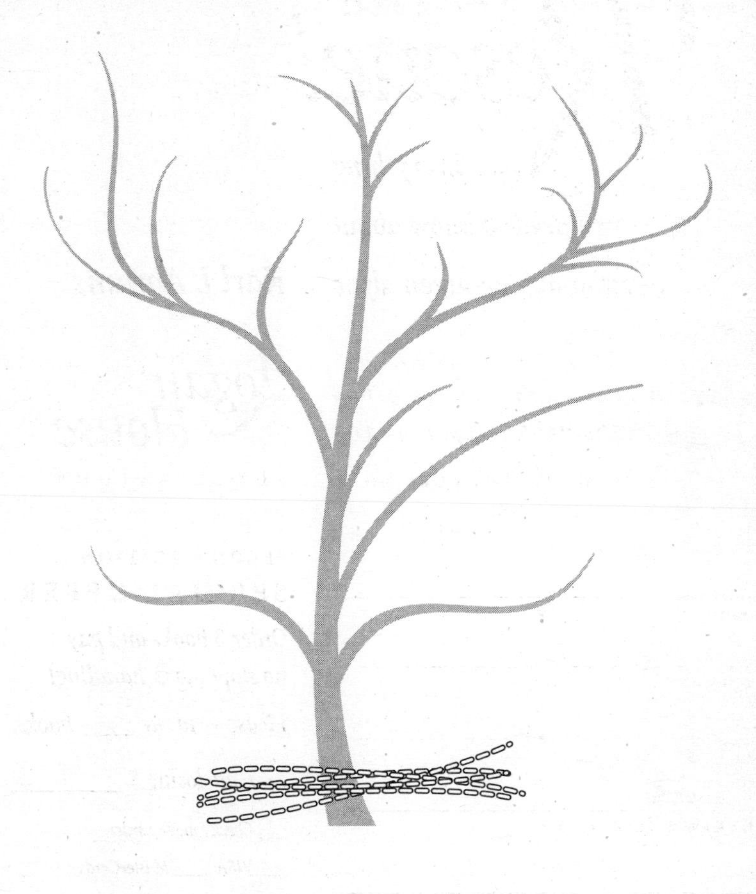

To keep the body in good health is a duty . . .
otherwise we shall not be able to keep
the mind strong and clear.

Gautama Buddha